MODERN SCIENTIFIC TOOLS IN BIOPROCESSING

DEVELOPMENTS IN WATER SCIENCE, 49

OTHER TITLES IN THIS SERIES

4 J.J. FRIED
GROUNDWATER POLLUTION
13 M.N. MARIÑO AND J.N. LUTIN
SEEPAGE AND GROUNDWATER
14 D. STEPHENSON
STORMWATER HYDROLOGY AND DRAINAGE
18 J. BALEK
HYDROLOGY AND WATER RESOURCES IN TROPICAL REGIONS
19 D. STEPHENSON
PIPEFLOW ANALYSIS
22 H.C. RIGGS
STREAMFLOW CHARACTERISTICS
25 W. KINZELBACH
GROUNDWATER MODELLING
26 D. STEPHENSON AND M.E. MEADOWS
KINEMATIC HYDROLOGY AND MODELLING
28 M. JERMAR
WATER RESOURCES AND WATER MANAGEMENT
29 G.W. ANNANDALE
RESERVOIR SEDIMENTATION
30 D. CLARKE
MICROCOMPUTER PROGRAMS IN GROUNDWATER
31 R.H. FRENCH
HYDRAULIC PROCESSES ON ALLUVIAL FANS
32 L. VOTRUBA, Z. KOS, K. NACHÁZEL, A. PATERA AND V. ZEMAN
ANALYSIS OF WATER RESOURCE SYSTEMS
33 L. VOTRUBA AND V. BROŽA
WATER MANAGEMENT IN RESERVOIRS
34 D. STEPHENSON
WATER AND WASTEWATER SYSTEMS ANALYSIS
35 M.A. CELIA ET AL. (EDITORS)
COMPUTATIONAL METHODS IN WATER RESOURCES, 1, MODELLING SURFACE AND SUB-SURFACE FLOWS
36 M.A. CELIA ET AL. (EDITORS)
COMPUTATIONAL METHODS IN WATER RESOURCES, 2, NUMERICAL METHODS FOR TRANSPORT AND HYDROLOGICAL PROCESSES
37 D. CLARKE
GROUNDWATER DISCHARGE TEST SIMULATION AND ANALYSIS MICROCOMPUTER PROGRAMMES IN TURBO PASCAL
38 J. BALEK
GROUNDWATER RESOURCES ASSESSMENT
39 E. CUSTODIO AND A. GURGUI (EDITORS)
GROUNDWATER ECONOMICS
40 D. STEPHENSON
PIPELINE DESIGN FOR WATER ENGINEERS
third revised and updated edition
41 D. STEPHENSON AND M.S. PETERSON
WATER RESOURCES DEVELOPMENT IN DEVELOPING COUNTRIES
42 K. NACHÁZEL
ESTIMATION THEORY IN HYDROLOGY AND WATER SYSTEMS
43 J. ZÁRUBA
WATER HAMMER IN PIPE-LINE SYSTEMS
44 W.E. KELLY AND S.MAREŠ (EDITORS)
APPLIED GEOPHYSICS IN HYDROGEOLOGICAL
45 K.W. HIPEL AND A.I. McLEOD
TIME SERIES MODELLING OF WATER RESOURCES AND ENVIROMENTAL SYSTEMS
46 G.A. BRUGGEMAN
ANALYTICAL SOLUTIONS OF GEOHYDROLOGICAL PROBLEMS

MODERN SCIENTIFIC TOOLS IN BIOPROCESSING

Reprinted from *Water Research,* Volume 36/2

P. WILDERER

Technische Universität, München, Am Coulombwall, D-85748, Garching, Germany

S. WUERTZ

University of California Davis, Department of Civil and Environmental Engineering, Davis, CA 95616, USA

2002
ELSEVIER
Amsterdam - Boston - London - New York - Oxford - Paris
San Diego - San Francisco - Singapore - Sydney - Tokyo

ELSEVIER SCIENCE B.V.
Sara Burgerhartstraat 25
P.O. Box 211, 1000 AE Amsterdam, The Netherlands

First edition 2002
Reprinted from Water Research 36/2
The reference edition of Modern Scientific Tools in Bioprocessing is a compilation of the archival material reprinted from the regular monthly edition. For citation purposes, the page numbers of reprinted articles are common to both editions.

Library of Congress Cataloging in Publication Data
A catalog record from the Library of Congress has been applied for.

ISBN: 0-444-51006-0

∞ The paper used in this publication meets the requirements of ANSI/NISO Z39.48-1992 (Permanence of Paper). Printed in The Netherlands.

CONTENTS

Special Issue: Modern Scientific Tools in Bioprocessing

Guest Editor: P. Wilderer

Peter Wilderer and Stefan Wuertz	369	Editorial
Peter A. Wilderer, Hans-Joachim Bungartz, Hilde Lemmer, Michael Wagner, Jurg Keller and Stefan Wuertz	370	Modern scientific methods and their potential in wastewater science and technology
R. Witzig, W. Manz, S. Rosenberger, U. Krüger, M. Kraume and U. Szewzyk	394	Microbiological aspects of a bioreactor with submerged membranes for aerobic treatment of municipal wastewater
Johwan Ahn, Tomotaka Daidou, Satoshi Tsuneda and Akira Hirata	403	Characterization of denitrifying phosphate-accumulating organisms cultivated under different electron acceptor conditions using polymerase chain reaction-denaturing gradient gel electrophoresis assay
S. Rosenberger, U. Krüger, R. Witzig, W. Manz, U. Szewzyk and M. Kraume	413	Performance of a bioreactor with submerged membranes for aerobic treatment of municipal waste water
Jeppe Lund nielsen and Per Halkjær Nielsen	421	Enumeration of acetate-consuming bacteria by microautoradiography under oxygen and nitrate respiring conditions in activated sludge
Futoshi Kurisu, Hiroyasu Satoh, Takashi Mino and Tomonori Matsuo	429	Microbial community analysis of thermophilic contact oxidation process by using ribosomal RNA approaches and the quinone profile method
G. L. Cronje, A. O. Beeharry, M. C. Wentzel and G. A. Ekama	439	Active biomass in activated sludge mixed liquor
Francis L. de los Reyes III and Lutgarde Raskin	445	Role of filamentous microorganisms in activated sludge foaming: relationship of mycolata levels to foaming initiation and stability
Giuliano Ziglio, Gianni Andreottola, Silvia Barbesti, Giorgio Boschetti, Laura Bruni, Paola Foladori and Roberta Villa	460	Assessment of activated sludge viability with flow cytometry
Regina Nogueira, Luís F. Melo, Ulrike Purkhold, Stefan Wuertz and Michael Wagner	469	Nitrifying and heterotrophic population dynamics in biofilm reactors: effects of hydraulic retention time and the presence of organic carbon
Zhiguo Yuan and Linda L. Blackall	482	Sludge population optimisation: a new dimension for the control of biological wastewater treatment systems
Christina M. Falkentoft, Elisabeth Müller, Patrik Arnz, Poul Harremoës, Hans Mosbæk, Peter A.Wilderer and Stefan Wuertz	491	Population changes in a biofilm reactor for phosphorus removal as evidenced by the use of FISH
A. Gieseke, P. Arnz, R. Amann and A. Schramm	501	Simultaneous P and N removal in a sequencing batch biofilm reactor: insights from reactor- and microscale investigations

PERGAMON

Water Research 36 (2002) 369

WATER RESEARCH

www.elsevier.com/locate/watres

Editorial

In recent years, a great variety of novel analytical methods has been developed to analyze composition, architecture and physico-chemical properties of microbial aggregates such as activated sludge flocs and biofilms. Simultaneously, new modeling approaches, aided by improved numerical simulation of the structure and function of these aggregates, have furthered our ability to understand their development and internal organization. The application of these novel analytical tools has led to fascinating discoveries, but has created a great deal of confusion in the engineering community as well.

With this issue of Water Research we want to help bridge the gap that emerged between science and engineering in the field of advanced biological wastewater treatment. Information is provided about methods which became available in recent years, both in microbiology and computer based modeling and simulation. Various authors elucidate the essence of the newly developed methods, the potentials these methods have in gaining better understanding of complex microbial systems, and the advantages which are envisioned with respect to optimization of biological wastewater treatment plants, trouble shooting and innovation.

Modern methods of molecular microbiology have gained tremendous popularity. However, correlation between results of microbial analysis and plant performance has still not been well established. The scientific and engineering community is encouraged to report not only on analytical results but also provide information about the history of analyzed samples and the performance of the plant under examination. The questionnaire enclosed in this issue should remind everybody that a complete data set is essential for engineers to be able to sufficiently exploit the information pro vided by modern microbiology. The questionnaire may be downloaded from the following web page: www.wga.bv.tum.de (click on either "data base" or "Daten bank"). Users of the questionnaire are invited to send back a copy of the data to be incorporated into a data bank. The content of this data bank will be available to the public free of charge.

Peter Wilderer
Institute of Water Quality Control and Waste Management, TU Munich, Am Coulombwall, 85748 Garching, Germany
E-mail address: p.wilderer@bv.tu-muenchen.de

Stefan Wuertz
University of California Davis, Department of Civil and Environmental Engineering, Davis, CA 95616, USA

PII: S0043-1354(01)00219-6

PERGAMON

Water Research 36 (2002) 370–393

WATER RESEARCH

www.elsevier.com/locate/watres

Modern scientific methods and their potential in wastewater science and technology

Peter A. Wilderer[a,*], Hans-Joachim Bungartz[b], Hilde Lemmer[c], Michael Wagner[d], Jurg Keller[e], Stefan Wuertz[a]

[a] *Institute of Water Quality Control and Waste Management, Technical University of Munich, Am Coulombwall, D-85748 Garching, Germany*

[b] *Institute of Mathematics, University of Augsburg, Universitätsstr. 14, D-86135 Augsburg, Germany*

[c] *Bavarian State Office for Water Management, Kaulbachstr. 37, 80539 München, Germany*

[d] *Department of Microbiology, Technical University of Munich, Am Hochanger 4, D-85350 Freising, Germany*

[e] *Advanced Wastewater Management Centre, University of Queensland, Brisbane, 4072, Australia*

Received 18 July 2000; received in revised form 2 April 2001; accepted 17 April 2001

Abstract

Application of novel analytical and investigative methods such as fluorescence in situ hybridization, confocal laser scanning microscopy (CLSM), microelectrodes and advanced numerical simulation has led to new insights into micro- and macroscopic processes in bioreactors. However, the question is still open whether or not these new findings and the subsequent gain of knowledge are of significant practical relevance and if so, where and how. To find suitable answers it is necessary for engineers to know what can be expected by applying these modern analytical tools. Similarly, scientists could benefit significantly from an intensive dialogue with engineers in order to find out about practical problems and conditions existing in wastewater treatment systems. In this paper, an attempt is made to help bridge the gap between science and engineering in biological wastewater treatment. We provide an overview of recently developed methods in microbiology and in mathematical modeling and numerical simulation. A questionnaire is presented which may help generate a platform from which further technical and scientific developments can be accomplished. Both the paper and the questionnaire are aimed at encouraging scientists and engineers to enter into an intensive, mutually beneficial dialogue.

Keywords: Wastewater treatment; FISH; Gene probes; Nucleic acid fingerprinting; Mathematical modelling; Numerical simulation

1. Introduction

In recent years, a great variety of analytical and investigative methods have been developed to analyze species composition, spatial structure (or architecture) and functional properties of microbial aggregates such as activated sludge flocs and biofilms. Simultaneously, novel methods became available which allow advanced mathematical modeling and numerical simulation of biological, chemical and physical processes in bio-aggregates and bioreactors. Application of these innovative tools led to scientific insights, but to some confusion as well. It has become apparent in the light of new findings that a considerable fraction of textbook knowledge needs to be rewritten. Practitioners, on the other hand, argue that the existing activated sludge and biofilm reactors work properly, disregarding all these new findings. They question whether it is necessary to know about all the details on the micro-scale when

*Corresponding author. Tel.: +49-89-289-13700; fax: +49-89-289-13718.

E-mail address: p.wilderer@bv.tu-muenchen.de (P.A. Wilderer).

0043-1354/02/$ - see front matter
PII: S0043-1354(01)00220-2

designing and operating large-scale treatment plants. It is far from being accepted that the increased level of analytical resolution in time and space satisfies the needs of the wastewater engineer. Certainly, shedding light on the black box of microbial systems is important for any further development in biotechnology. Yet does a colorful photograph obtained with a confocal laser scanning microscope (CLSM) and analyzed by digital image analysis provide any information of practical value? Is it important for the engineer to know the phylogenetic position of functionally important bacterial species in wastewater treatment plants? For example, does it matter whether the ammonia oxidizers active in an activated sludge plant are to be called *Nitrosomonas europaea* or *Nitrosococcus mobilis*, as long as the treatment plant functions as it should? Is it necessary to know about metabolic function on the cellular level, when a reactor of several hundreds of cubic meter volume needs to be designed and operated? Is it wise to advocate replacement of the empirical rule-of-thumb design procedures by mathematical models? Current models contain a vast number of default process parameter values. It can be argued that it is better to analytically determine as many parameters as possible and incorporate concrete numbers into models instead of fitting them with an estimated factor.

Many of the current design and operating routines have very little, if any fundamental support from a scientific point of view. Why should the aerobic treatment processes always run at a DO setpoint of $2\,mg\,l^{-1}$? What is the reason behind the differences in the maximal specific growth rates for nitrifiers between different treatment plants? Given that this is one of the most critical design parameters in biological nutrient removal (BNR) plant design, it would be worthwhile to know why there are differences and how they could possibly be predicted or overcome. Why does the biological P removal performance vary drastically in some plants despite seemingly suitable wastewater compositions? Particularly for optimization and troubleshooting at a practical level, a more scientific rather than empirical (trial-and-error) approach could be highly valuable in many cases.

The underlying reason behind many of these questions is often a lack of proper information on both sides, the practitioners and the scientists. The consulting engineers have lost track of the developments that take place in the laboratories, and the scientists lack experience of real-world systems. However, it is most important to overcome the confusion that developed over the past few years, and to bridge the gap between science and engineering in the field of wastewater treatment technology. If this can be achieved, significant practical benefits and advances will be possible, increasing the awareness and valuation of each other's knowledge and expertise. This paper is attempting to provide assistance in the process of creating a dialogue and matching the interests of both sides. In the following, information is given about the principles and potential of the novel methods currently available or under development. We try to summarize how the various analytical methods work and which type of practically relevant information can be deduced. We also introduce a questionnaire to be filled in when subjecting a sample to microbial analysis. It contains the most important information about the history of the sample and the methods applied to evaluate its population structure and function. Correlation of these historical data with the results obtained by the novel methods should provide the basis for the development of a data bank which can serve as a commonly available information basis for any further technology development and/or for troubleshooting at existing treatment plants.

2. Description of novel analytical methods and their applications

In the following, the term "microbial aggregate" refers to associations of microorganisms growing either as activated sludge flocs or as biofilms. These aggregates, which are responsible for the removal of pollutants, consist of a wide variety of often not yet culturable bacterial species, protozoa and metazoa, possibly also of fungi and yeast cells. The organisms feed either on dissolved or on particulate material. Organisms capable of performing aerobic metabolism are present as well as organisms, which are active under anoxic and anaerobic conditions. Some of the organisms belong physiologically to the group of chemolithoautotrophs, others to the group of chemoorganoheterotrophs. Even phototrophic organisms can be found. With respect to activated sludge flocs, organisms producing extracellular polymeric substances (EPS) as major floc components are important. They may attach to certain filamentous organisms believed to contribute to floc stability. The settleability of sludge depends greatly on the formation of compact flocs with a surface charge distribution allowing coagulation of larger aggregates, which settle quickly when hydraulic shear is reduced. Microorganisms, which use the flocs as a support but grow in the form of filaments into the surrounding water (as opposed to stabilizing the floc), are very undesirable in the process since under reduced shear conditions they flocculate poorly, producing voluminous aggregates of very low density (bulking sludge). These aggregates do not settle well and may—in the worst case—cause activated sludge plants to fail due to the washout of sludge with the effluent.

The diversity of microbial species in an activated sludge plant or a biofilm reactor is the result of the composition of the influent wastewater, environmental parameters such as pH and temperature, and the process conditions prevailing in the reactor or plant. Of importance are factors such as mean solids retention time (sludge age), hydraulic retention time distribution, oxygen and substrate (pollutant) concentrations, and periodicity provided by the recycling of sludge or by the cycle frequency in sequencing batch reactors. Some of the factors are purposely selected by the engineer in order to affect the species composition and by that the overall metabolic activity of the microbial community. Others must be taken as they appear (e.g., wastewater composition) or are simply not recognized as relevant factors.

Modern analytical methods as they were developed, for instance, in molecular microbiology can be applied to investigate the relationship between species composition of microbial aggregates and the prevailing process parameters. Knowing about the correlation between these two categories of factors may allow the engineer to better control the microbial system in wastewater treatment plants, and to select tailored process strategies during start-up and operation of bioreactors. An overview of cultivation-independent molecular biological, chemical and physical methods currently available to investigate microbial aggregates is given below and summarized in Table 1.

3. Microbial population analysis

3.1. Environmental 16S and 23S rRNA libraries and FISH

In some instances it may be necessary to obtain a complete inventory of the microorganisms in a treatment plant or reactor. Since most of the bacterial cells will not grow on standard laboratory media, an alternate approach is to extract their DNA and to construct a data bank consisting of specific genes, which can be used to identify bacterial cells independent of cultivation. These 16S and 23S rDNA genes code for the rRNA molecules needed for protein synthesis; they are about 1500 and 3000 nucleotides in length, respectively, and contain both highly variable and conserved regions. Extracted 16S or 23S rDNA (or rRNA, which are first converted to complimentary DNA strands using the enzyme reverse transcriptase), are amplified enzymatically using the polymerase chain reaction (PCR) [13]. By employing specific primers (i.e. oligonucleotides which bind initially to the target DNA to be amplified) only 16S or 23S rDNA sequences are amplified which can then be cloned and maintained in *Escherichia coli* serving as host and further multiplying the DNA copies as part of its growth cycle. These *E. coli* cells effectively act as a gene library, an inventory of rRNA of the original microbial population. The DNA fragments can be sequenced to reveal the identity of the corresponding bacteria by comparative analysis with 16S rDNA sequences of other bacteria. Currently, more than 15,000 16S rRNA sequences are maintained in public data bases. The 23S rRNA data base is still limited in size but is expected to grow rapidly in the future. There are limitations of the ribosomal DNA approach due to potential biases introduced during nucleic acid extraction from the wastewater sample, DNA amplification and cloning in *E. coli* and the fact that the bacterial chromosome can have up to 15 copies of the rRNA genes. The net result is that an environmental rRNA library can never be a quantitative estimate of microbial abundance. Rather, it forms the basis of what we term the full cycle rRNA approach (Fig. 1). It is useful to understand its underlying principle in order to apply gene probes for process monitoring and optimization as outlined below.

16S rDNA sequences have been instrumental in classifying bacteria and in assigning them their position in phylogenetic trees. Although the latter is not always relevant to the practicing engineer there are several useful applications, which stem from increased knowledge in the area of phylogeny. Gene probes (single-stranded sequences of DNA generally about 17–21 bases in length) can be designed which are specific for the rRNA of certain bacteria and will hybridize to them, that is, form double-stranded structures. Hybridization can occur ex situ to isolated nucleic acids (for example, on a membrane such as in dot-blot hybridization) or in situ to whole cells. By attaching a fluorophore to the gene probe and subsequent microscopic detection (a method known as fluorescent in situ hybridization (FISH)) it is possible to count cells of particular interest and draw conclusions based on their presence if their physiological role in wastewater treatment is known (for example, nitrifying bacteria). Both methods, dot-blot hybridization and FISH, are quantitative and target rRNA which is present mainly in active cells. One limitation of ex situ nucleic acid hybridization is that the relative abundance of rRNA cannot be directly translated into cell numbers since ribosome contents of different bacterial species vary from about 10^3 to 10^5 ribosomes per cell and are a function of growth rate in many species. Consequently, whole cell hybridization using FISH is the preferred method to estimate bacterial cell numbers if knowledge of their identity is desired. In fact, FISH is the ultimate method of choice to verify the presence and in general terms, the cellular activity based on the fact that rRNA is degraded, albeit slowly, in inactive or dead cells of microorganisms in activated sludge or biofilms. The main advantage is that it does not involve any DNA or RNA amplification but allows microscopic inspection of intact cells in samples (Fig. 1). It can be performed with relatively simple equipment and no extensive training of laboratory personnel

Table 1
Comparison of high resolution methods for investigation of microbial aggregates

Method	Suitable for[a]	Advantages	Limitations	Reference examples
Environmental 16S rRNA and functional gene libraries	D, PS	High phylogenetic resolution; required to design new gene probes and to identify novel bacteria	Time consuming; DNA extraction, PCR and cloning biases; not quantitative in regard to community composition	[1,2,21]
16S rRNA and functional gene fingerprinting techniques	D, PS	Inexpensive; high sample throughput	DNA extraction and PCR biases; not quantitative in regard to community composition; multiple rRNA operons per organism can complicate analysis	[3–5]
FISH	D, PS, A	Relatively straightforward to perform; allows direct visualization of non-cultured microbes; amenable to direct combination with microautoradiography and microsensors	Adaptation of protocols required for some gram-positive *Bacteria* and for *Archaea*; inactive cells with low ribosome content not detectable; sample embedding required to preserve aggregate architecture	[1,6,7,17]
FISH/fluorescent staining and CLSM and image analysis	A, D, PS	Quantitative analysis of population structure and/or floc architecture; non-invasive technique; amenable to automation	Expensive instrumentation; time consuming	[8,9]
FISH—Microautoradiography	D, PS, MA	Simultaneous in situ identification and determination of substrate uptake pattern on a single cell level; quantification of important physiological microbial groups (e.g. denitrifiers) possible	Specialized knowledge required; isotope lab needed; time consuming and expensive	[10]
FISH—Microsensors	D, PS MA	Chemical concentration gradients can be directly measured and correlated with spatial arrangement of microbial consortia	Invasive technique; usually one-dimensional measurements: problems might arise due to architectural heterogeneity of aggregates; spatial resolution not on single microbial cell level; not all compounds measurable; possible interference by other ions	[11,12]

[a] Estimation of microbial diversity (D); determination of microbial population structure and dynamics (PS), aggregate architecture (A), specific microbial activities (MA), chemical concentration gradients (C).

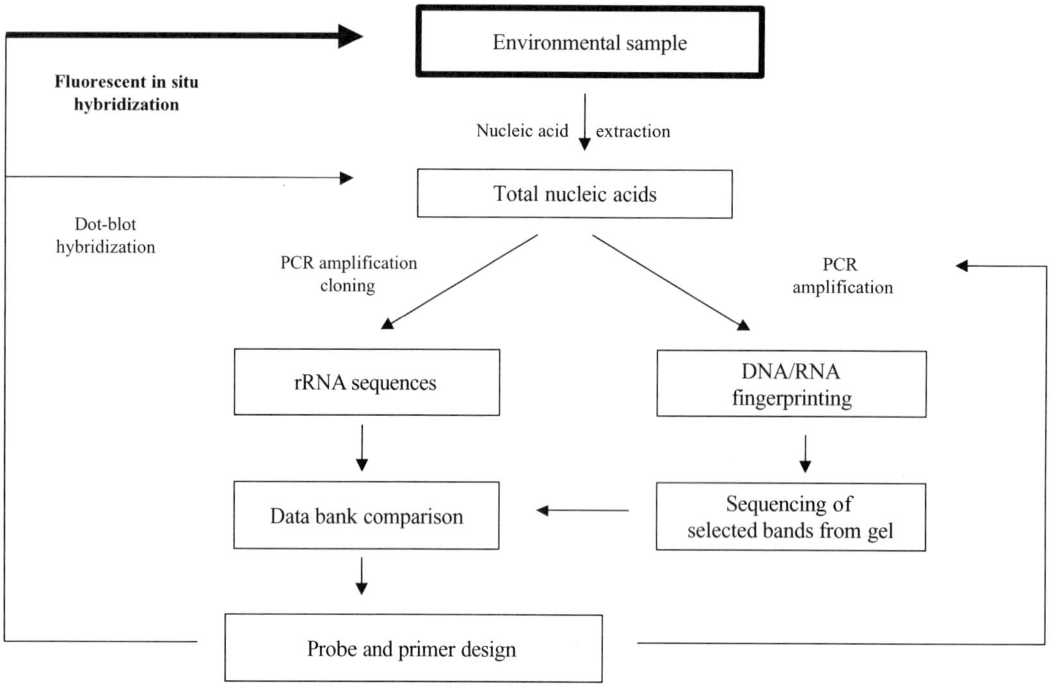

Fig. 1. Flow chart for full cycle rRNA analysis. FISH shown in boldface is recommended for routine analysis by testing laboratories and plant operators. The other steps in the cycle are necessary to obtain a complete picture of the microbial populations in a reactor. They can be performed by testing laboratories to obtain supplementary information regarding the diversity of bacteria present (using fingerprinting methods) and to construct nucleic acid probes (cloning of rRNA sequences, construction of a gene library and design of new gene probes) for more specific FISH analysis as needed.

beyond microscopic expertise is required (Table 1). The involvement of an experienced consultant is recommended for decisions regarding the choice of gene probes and the interpretation of FISH results. The ability to identify key organisms microscopically is useful, for example, as an early warning system for filamentous bulking and scum events [14–16] or to monitor the functioning of nitrification plants (Fig. 2) [2,12,17,18]. FISH analysis in combination with environmental 16S rRNA libraries has contributed to the discovery and understanding of processes and microorganisms relevant in wastewater treatment which were hitherto unknown. Examples of this include the anaerobic oxidation of ammonium (ANAMMOX) [19,20], polyphosphate accumulating organisms (PAO) responsible for biological P removal [21] or various nitrification organisms such as *Nitrospira* [17,22].

3.2. Nucleic acid fingerprinting

Denaturing gradient gel electrophoresis (DGGE) is one of several DNA/RNA fingerprinting methods used to investigate microbial communities including single-strand conformation polymorphism analysis (for example, [4]), temperature gradient gel electrophoresis [23], and terminal restriction length polymorphism [5]. DGGE will be discussed here as an example of information that can be gained and limitations to be considered. In recent years the use of fingerprinting methods has become popular in molecular microbial ecology as an alternative to the more laborious analysis of 16S rRNA gene libraries [24]. Like other molecular biological methods it is based on the amplification of specific nucleic acid fragments by the PCR and analysis by gel electrophoresis. However, the size of the amplified DNA fragments is limited to about 500 bp, which severely hampers subsequent probe design and phylogenetic analyses (see below). In contrast to conventional electrophoresis that allows the separation of DNA fragments based on their size, DGGE enables the separation of DNA fragments of identical length but different sequence.

Thus, microorganisms differing in their 16S rRNA genes will produce unique gel bands. Different assemblages of microbial species consequently lead to different patterns, which alone do not give any information about the identity of the bacterial species present. DGGE has been applied to study microbial community complexity (for example, [23,25,26]), to observe population shifts (for example, [27]), to follow the expression of relevant

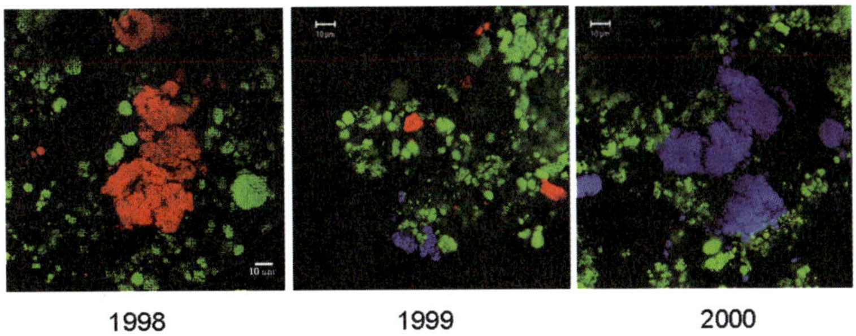

Fig. 2. FISH analysis of ammonia- (labelled in green), nitrite- (labelled in red) and anaerobic ammonia-oxidizers (labelled in blue) in a sequencing batch biofilm reactor. After 1998 anaerobic phases were introduced in the cycling mode which led to a replacement of the "aerobic" nitrite oxidizers by the anaerobic ammonium oxidizers (CLSM images recorded by Markus Schmid).

genes in the environment [28], and to check the efficiency of enrichment cultures for the isolation of bacteria (for example, [26,29]). This is a powerful method well suited to demonstrate the diversity of a particular wastewater community. For example, the diversity of a microbial population in different reactor types receiving the same wastewater input can be analyzed by comparing fingerprinting patterns. Bands of interest can be eluted from gels and sequenced to identify the corresponding bacterial species. By designing a FISH probe based on the eluted and cloned sequence the actual abundance of a gel band that appears to be dominant after DGGE analysis can be verified in situ. To this end the original wastewater sample is hybridized with the probe and analyzed by epifluorescence microscopy or CLSM. Investigators should bear in mind that only partial sequences are available from DGGE bands (see above).

If one is not interested in general microbial diversity, but rather in the occurrence of organisms capable of performing particular metabolic functions such as N transformations or degradation of polycyclic aromatic hydrocarbons, the choice of primer for the PCR amplification step can be narrowed to include only a particular group of organisms or the expression of a specific functional gene. The scope of this type of analysis is considerable as it allows the verification of the presence, e.g. of *Nitrosomonas* or *Nitrobacter* species, or the expression of ammonium monooxogenase. To get an indication of the active members of the community it is advisable to target rRNA (for the identification of organisms) and mRNA (for the detection of gene expression) rather than DNA, the latter being present extracellularly and in inactive or dead cells also.

DGGE has been used on a deteriorated reactor for enhanced biological phosphorus removal (EBPR) in combination with FISH to first detect dominant 16S rDNA bands on denaturing gradient gels and then design FISH probes based on the sequences of these bands [3]. FISH analysis showed that 35% of the microbial population consisted of this group of bacteria compared to an estimated 75% based on DGGE band intensities. Hence DGGE can be used to short-cut the more laborious clone bank approach but it should always be verified by whole cell hybridization analysis (Fig. 1).

To summarize, DGGE and other fingerprinting techniques are ideally suited for high-throughput screening of reactors. Their main use lies in the comparison of samples differing only in the effect of one or more variables such as time or operating conditions. By varying the primer sets in the PCR amplification the desired specificity of the fingerprinting pattern can be adjusted. However, compared to FISH of whole cells it is subject to numerous biases pertaining to DNA amplification and cloning and thus does not produce quantitative results.

4. Structure-function analysis

4.1. Confocal laser scanning microscopy and image analysis

Microscopic cell counts can be tedious and tiring and are sometimes prone to subjectivity. To be widely useful as a monitoring tool the process of image acquisition must be automated and standardized. In contrast to conventional optical microscopy, CLSM allows the observer to remove out of focus information and perform depth-resolved scanning. The sample does not have to be pretreated and can be inspected in its natural hydrated state. Thus CLSM has become the method of choice for the analysis of flocs and biofilms in reactors. The principle involves excitation of a fluorophore (e.g., a fluorescent dye binding to nucleic acids or other target structures in the cell) by a focused laser beam. Once an image has been digitized it can be analyzed by image processing to derive additional information. Image analysis can be quantitative [8,9]. For example, bacterial cell numbers have been determined based on fluorescence signals after nucleic acid staining [30–33]. Similarly, one can quantify microbial areal fractions or cell volumes in bioaggregates [20,34]. The determination of cell numbers is difficult due to the great heterogeneity of samples and the spatial architecture involved. In our hands, automated cell counting has proved successful only for homogenized single cell layers.

Hence, two approaches are recommended. Areal fractions of cells in relatively tightly packed flocs responding to a specific fluorescent tag can be analyzed to obtain information about the relative frequency of microbial populations identified by FISH. The same procedure can be applied to biofilms if they are fairly homogeneous. However, due to the presence of a substratum which does not allow access of nutrients from all directions as in flocs, the zonation of biofilms in terms of cell density, areal porosity, identity of microorganisms, nutrient and oxygen supply among others tends to be greater. Consequently, the spatial variability may be such that biovolumes (involving several optical planes) rather than areal fractions should be used to analyze microscopic data [8].

A relatively simple approach to the calculation of biovolumes based on numerical integration of CLSM-derived binary images has been developed, which can be automated and incorporated into commercially available software [8]. The advantage of automated microscopy and image processing is that large sections of biofilm or activated sludge can be scanned. To date, few attempts have been made to investigate statistical treatments of thus derived microscopic images [35–37]. In recent years quantitative image analysis of microbial aggregates has burgeoned [37–39] and there are now methods which allow automated threshold selection during image acquisition [37] thus greatly simplifying microscope handling. The increasing availability of these analytical tools is expected to accelerate the general utility of FISH and light microscopy in general for the rapid screening of treatment plants. An area of active research is the automated detection and computation of filament lengths and its correlation with bulking events.

4.2. FISH and microautoradiography

The microscopic detection of microbial cells and the ability to estimate their abundance in reactors provide valuable information about population structure, but lack proof that the organisms in question are actually performing a certain function. Recently, FISH has been combined with microautoradiography (MAR) to determine specific uptake of organic and inorganic radiolabeled substrates by activated sludge microbial communities [10]. This method can be used, for example, to verify that certain organisms, which are purported to perform biological P removal, are indeed active during periodic changes between anoxic/aerobic and anaerobic conditions. FISH and MAR have been used to demonstrate that members of the nitrite oxidizing genus *Nitrospira* in activated sludge, which could not be cultivated in the laboratory, are able to fix atmospheric CO_2. They were far more abundant in situ than were *Nitrobacter* cells [40]. This type of knowledge is valuable when it comes to improving mathematical models and understanding why certain physiological conditions must be optimized in order for nitrification to persist in a particular reactor setting.

4.3. Microsensors

Microsensors are useful tools to study chemical concentration gradients and microbial activities in biofilms and flocs [41]. To date there are devices for a number of important substances including O_2 [42], CH_4, CO_2, S^{2-}, H_2S [43], NO_3^- [44], NO_2^-, N_2O [45], and NH_4 [46] as well as pH [47]. They are needle shaped sensors with a tip size of 1–20 μm and can be subdivided according to the principle of operation.

Amperometric sensors (H_2S, HClO, N_2O, O_2) measure currents based on reduction or oxidation of a substrate inside the tip; potentiometric sensors record the electric potential at the tip generated by charge separation and optrodes are optical fibers with a fluorescent indicator dye covering the tip. Their general applications and preparation have been described elsewhere [41,48]. Amperometric sensors such as the O_2 microelectrode have a tip diameter of 2–10 μm. Among the most promising potentiometric sensors are those based on liquid ion-exchange membranes (LIX). With a tip diameter of 1 μm or less they are useful for in situ studies of biofilms and activated sludge flocs. Examples of LIX sensors are Ca^{2+} [49], CO_2, CO_3^{2-} [50], H^+ [51], NH_4^+ [46], NO_3^- [44] and NO_2^-. Optrodes, [41,52] which are much easier to prepare, have a tip diameter of at least 20 μm and although they are at present too big to probe microbial aggregates, they may still be useful for measurements of O_2, pH and temperature on a microscale. The power of microsensors lies in their ability to measure local concentration gradients directly. Although invasive, their tip diameter, if small enough, will not greatly disturb the architecture of microbial aggregates. This spatial resolution, when related to the presence of microbial communities and to the bulk fluid parameters (concentration, shear, etc.), can lead to important new insights into the structure and function of aggregates. For example, one may wish to determine the distribution of ammonia oxidizers and nitrite oxidizers by applying FISH to samples from a biofilm reactor. However, without any knowledge about the actual concentration profiles of NO_2^-, NO_3^- and NH_4^+ no statements regarding the activities of these microbial populations can be made. With the help of microsensors the investigator can determine effective diffusion coefficients (D_{eff}) inside aggregates and calculate diffusional transport of substrates [41]. For comparison, D_{eff} can also be determined by following fluorescent tracers using CLSM [53].

4.4. CLSM, FISH and microsensors

Studies where microsensor techniques have been applied to wastewater systems are only emerging slowly [11,12,54,55]. Schramm et al. [12] reported that in a fluidized bed reactor NH_4^+ and O_2 were present at depths of 10 μm and 70 μm of the biofilm formed, respectively. By applying FISH techniques the authors could demonstrate dense populations of nitrifiers deeper than 125 μm, which were apparently inactive. Without the application of microsensors one could easily have concluded that nitrifying activities extended all the way to the substratum. Hence, through the combination of two novel techniques, microsensors and FISH, significant information about the structure and function of microbial aggregates can be gained.

CLSM in combination with FISH is a powerful approach to investigate microbial community structure and aggregate architecture. The recognition that biofilms do not always represent homogeneous structures was largely driven by the use of CLSM and microelectrodes. The former technique revealed pores and channels in certain biofilms, whereas the latter resolved concentration profiles inside of biofilms. It is now accepted that mass transport is not necessarily diffusion limited. Rather, the architecture of the microbial aggregate determines whether convective transport can occur. Obviously, the hydraulic regime plays a great role in such a case.

One limitation of CLSM and microelectrodes, although in the opposite sense, is their depth of penetration. Light microscopy is limited to a thickness of about 300 μm. Most biofilms that have been investigated by CLSM were less than 100 μm thick. In wastewater reactors, biofilms can vary in thickness from 0.1 to several millimeters. Hence, engineers have been reluctant to accept that biofilms in technical applications should allow convective transport throughout the biofilm depth. In order to answer this question biofilms must be sectioned (e.g. by cryosectioning) and preserved without damaging their integrity. Such prepared microslices are then amenable to analysis by CLSM. Microelectrodes, on the other hand, only work well in biofilms thicker than about 20 μm. This is due to the diameter of their tip, which can vary from 1 to 20 μm. In addition, the chances of damage or breakage of the microsensor increase with decreasing distance from the substratum.

To summarize, the use of CLSM and microelectrodes should be extended to include samples from pilot-scale and full-scale reactors. Biofilm systems including support media should be analyzed (e.g. by placing expanded clay beads or plastic materials in a petridish with a special cover slip to allow standard objective lenses to be used). If the biofilm is too thick it should be embedded to preserve structure integrity and sectioned. The information gained will tell the operator whether certain important bacterial key players are present in the reactor and throughout the microbial aggregates and whether the samples are diffusion limited. This is particularly important since a frequently held belief among engineers is that only the top few micrometers of a biofilm are metabolically active. Yet thick biofilms generally function well and loss of the microbial surface layer does not lead to a decrease in metabolic activity. In fact, deep biofilms can be particularly suitable for zoning whereby e.g. an aerobic surface layer can be used for nitrification while the underlying zone or layer might be anoxic, leading to denitrification. Such aggregates, either as biofilms or flocs, can have significant practical importance. However, control of the suitable reactor conditions to achieve and maintain the desired zones is crucial and relies heavily on the detailed knowledge of the layers and their activity. Analysis of the biofilm architecture would establish what metabolic activity is present in which layer.

4.5. Measures of microbial heterogeneity

Both the spatial architecture and the microbial population structure of microbial aggregates tend to be heterogeneous to varying degrees. Using CLSM or other methods yielding visual information one can quantify this heterogeneity as a prerequisite to a meaningful comparison of biofilm or floc structure and metabolic function. If such a relationship could be established it may bear relevance to the operation of a reactor system as a whole. Different approaches have been taken including an estimation of fractal dimension [56] or porosity [57]. The sine qua non of any attempt to quantity aggregate structure is that the features chosen must be related to the fundamental processes shaping the aggregate as manifested in optical sections taken by a CLSM [38]. To this end four parameters have been proposed which can be used to follow structural changes. They are: textural entropy as a measure of disorganization, areal porosity (defined as the ratio of void area to the total area of an image), fractal dimension, and maximum diffusion distance (defined as the maximum distance of cluster pixels to the nearest void pixel) [39]. Other approaches involve analysis of cell cluster size and distribution [93]. Efforts to correlate structural characteristics with substrate concentration profiles (as measured using microsensors) and the distribution of microbial populations (as detected using FISH) are underway in several laboratories. The benefits of these fundamental studies to reactor operation are only beginning to be explored.

5. Mathematical modeling of structure and function of flocs and biofilms

Due to the tremendous gain in performance of computer hardware and software seen in the last decades, the scope of the traditional discipline 'mathematical modeling' has widened dramatically. Today, the motivation for the derivation of a mathematical model, usually a system of algebraic or differential equations, is not only to get a formal description of the phenomena or processes to be modelled and, thus, to get insight into their laws. Rather, models are used as the starting point of numerical simulations, aiming at the prediction of some system's behavior or at the optimization of its function. Once the crucial quantities and their interactions are identified, and once numerical algorithms and the corresponding code to simulate those considered most important are available, parameter modifications, changes of initial or boundary conditions, or variations of the whole scenario can be realized—at significantly lower cost than required in changing experimental parameters. The typical steps in model development are:

- to identify the crucial quantities and their interactions;
- to define the numerical algorithms and the corresponding code to simulate those considered most important;
- to identify and calibrate the model parameters, initial and/or boundary conditions; and
- then to validate the model using actual experimental data.

Once a well validated model is available, numerous variations of the selected scenario can be simulated and studied, generally at significantly lower costs than the corresponding experimental studies.

Among all modern technological fields where simulation is applied, environmental bioreactors probably belong to the most challenging. This is mainly due to three reasons. First, there is a whole array of strongly interacting effects involved, ranging from purely physical (fluid dynamics, convective and diffusive transport, mechanical loads and detachment) over chemical (reactions, metabolism) to biological ones (coexistence of different species, behavior under competition). Second, there is an immensely wide range of relevant scales, with respect to both time and space. While the biochemical reactions occur in tiny "bioreactors", i.e., bacteria with a volume of a few femtoliters (10^{-15} l), the actual reactor tanks have often a volume of many megaliters (10^{6} l). Similarly, on a time-scale, some processes are very fast, e.g. hydraulic flow fields in a turbulent flow situation, while on the other hand, biomass growth is usually a very much slower process. Third, the various effects on the different scales require different mathematical methods for their description (ordinary differential equations, partial differential equations, stochastic processes, etc.) and, thus, different expertise for the respective numerical simulations.

Therefore, while it might be possible in principle to derive an exhaustive model that takes into account all of the relevant effects known so far, there is no chance of transferring such a model into one single computer program. Hence, restrictions and compromises regarding the effects to be studied and the scale of time and space are inevitable. This leads to simplifications and assumptions, which need to be carefully evaluated and assessed regarding their impact on the final solution. Furthermore, mechanisms have to be provided how the various simulation results can be combined, especially how the fine-scale results can be transferred to the coarse scale in order to improve the prediction there and meet the practitioner's interest. Although the underlying model is a crucial aspect of any numerical simulation, the success of the simulation depends on more factors. This is especially true if the simulation shall not be restricted to the use of simple commercially available packages, but if it includes the development of state-of-the-art algorithms and software, too. Hence, developments in different fields have extended the scope of simulations significantly over the last years, and many of these have made it into commercial packages over time.

In terms of merging engineering and technology of wastewater treatment systems with the new knowledge from microbiological and molecular biological analytical methods, it is necessary to provide as much quantitative information as possible. This data is crucial for the validation of more detailed, and particularly microscale models. It is quite easy today to develop a model of any process, however, only a model that has been shown to truly describe reality by comparing it to actual experimental data should be considered as a valuable tool for design and operation of treatment systems. The new information available from these novel tools will also allow the development of more specific and accurate models. For example, activated sludge models now in use worldwide for the design and operation of treatment systems use hypothetical concentrations of active organisms. Monitoring active organisms involved in nitrification, denitrification and biological P removal, that is, validate their concentrations and activities by actual measurement would be an advantage in the validation of the model. It is timely for the new microbiology and molecular biology techniques to provide suitable quantitative data to replace VSS as the "active biomass" measurement and to define specific COD, oxygen, and nutrient utilization rates among other kinetic information. This has been done for only a few systems such as nitrifying sludge [18]. Below, several facets of biofilm modeling are discussed as an example of the impact of the rapid development in simulation tools on our basic understanding of the biological processes in nutrient removal.

6. Aspects of modeling and numerical simulation

6.1. Mathematical models

Mathematical models are used to develop a mathematical description of existing or imaginary systems and processes. The key to modeling is to select relevant quantities and processes and describe them in a mathematical or logical format (equations, rules, transition probabilities, etc.). For biofilm systems, a broad spectrum of modeling approaches has been developed, such as continuous models with classical physical quantities, discrete models describing the transition from a given state to a successor state, stochastic models that are based on random processes, and so forth.

The engineer responsible for a biofilm reactor is primarily interested in its undisturbed, effective, and efficient operation. The relevant factors to achieve this in a biofilm reactor, and their interactions are not at all clear, however. This has led to a large number of modeling approaches, all with different trade-offs between accuracy and simplicity [58]. As outlined above, the scale of the model with respect to both time and space has to be defined. The key question in the model development phase is what the model should be used for finally. For example, do we want to describe the whole reactor or just a small section of the biofilm? The answers to these questions will have strong implications on the subsequent steps. Based on the observation that biofilm systems are essentially three-dimensional (3D), with a large variety of mostly heterogeneous morphological appearances [56] (which holds, at least to some extent, for nearly all scales), the next question is how to model the geometry in an adequate way. Should the model take into account full spatial resolution [59,60] or is a 1 or 2D layer model representative enough for the description of the relevant processes [61,62]? Or can even a homogeneous structure be assumed, which then allows to work with some global parameters like porosity only? The next step is to identify the factors affecting the system and select them to become part of the model. Some models are primarily addressing physical or physico-chemical effects like fluid flow, solute transport, chemical reactions, or mechanical stress [63]. Others, however, put the focus completely on the biological behavior in the sense of population dynamics [64], possibly even omitting the physical background.

Finally, depending on the above decisions, an appropriate mathematical apparatus has to be applied. In the physical case, this will usually be some system of differential equations (ordinary ones if geometry is neglected, partial ones if space is partially or fully resolved) [61,65]. To this group belong all models that are based on the principle laws of continuum mechanics [59] or models obtained from these via some homogenisation process [66]. In contrast to that, the variety of biological models is larger. Apart from equation-based models (here usually as ordinary differential equations), there are rule-based approaches that have been developed in order to represent paradigms like competition, growth, or movement. Here, the system of rules is based more on observations than any detailed measurements. Furthermore, the use of stochastic (individual) quantities is also common. Typical representatives of this group are models based on cellular automata [60] or fractals [56].

Recently, there have been attempts to bring together the physical and the biological world, or the equation- and the rule-based formalism [60]. Such hybrid models are more powerful with respect to the spectrum of effects they describe. On the other hand, their conversion into numerical simulations is not yet solved in a satisfactory way.

Although the operation of today's treatment plants is largely computer-based, the use of models for operation support is not yet very common. In process design, increasingly dynamic models are being applied but often only as an add-on to the traditional design guidelines. These models are generally all macroscopic, process oriented models. They do not take into account any geometric (e.g. flow patterns) or micro-scale spatial details. The potential of an explicit and detailed model of microscopic processes for further optimization is very significant. It is important to note that for spatially resolved simulations, the development of efficient algorithms and programs is still much more a research task itself than just an act of programming. Increasingly, commercial simulation software for solely time-dependent or simple layer-based biofilm models is becoming available [61,62]. This offers easy access to biofilm simulation tools, although, one has to keep in mind the trade-off between detail and accuracy on the one hand and simplicity on the other hand. In many cases only complete spatial resolution can provide insights into local details and heterogeneities as well as into their consequences.

6.2. Numerical algorithms and simulation software

Since the overwhelming majority of models cannot be solved analytically, all non-discrete model equations have to be discretized in order to be accessible for a computer, and the resulting large systems of equations have to be solved by some iterative scheme. These two aspects are the main contribution of numerical mathematics to any simulation. An efficient discretization helps to keep the number of unknowns moderate, and fast solvers reduce the computational effort (storage and computing time). This is crucial since the gain in processor speed (though impressive) is not enough to allow full-detail, realistic simulations. From the computational point of view, partial differential equations (PDE) are the most challenging part of present biofilm models. Actually, appropriate discretization techniques for PDE and the development of the corresponding supercomputer-suited algorithms and programs is still an active field of research. This holds especially for the simulation of the flow field and related phenomena like nutrient transport, chemical reactions, or detachment in complicated, heterogeneous, and varying biofilm geometries [60], which have entered the scope of numerical treatment only recently. Here, both classical techniques [59] (mainly finite differences or finite volumes) and the more recently developed Lattice Boltzmann methods [60,67] are used. For all of them strategies for a fast iterative solution of the resulting systems of equations have to be developed or adapted [68,69].

6.3. Tools for visualization and embedding

Since spatially resolved simulations are becoming increasingly common, the question of an appropriate and effective interpretation of their results is raised frequently. These results are no longer consisting of a few characteristic numbers, but encompass a whole field of scalar or vector quantities. Despite sometimes allergic reactions to 'colorful pictures' among engineers, sophisticated visualization tools are usually the best way to analyze and interpret the model outputs.

The possibility to generate spatially resolved 3D data offers a completely new challenge. So far, the results of microscopic investigations, X-ray examinations in medicine, aerial photographs, or numerical simulations had mostly resulted in characteristic numbers, lines or single images (2D data sets). Nowadays, in many of these fields, 3D data are available as a volume or as image stacks e.g. from computational fluid dynamic simulations or from CLSM images. Visualization deals with the graphical representation of such 3D data and often utilise sophisticated methods from image analysis or aspects from computer graphics. Application of visualization tools in biofilm research allows us to explore biofilm geometries (obtained from experiments or simulations) in their local details with the help of virtual flights through the sample, to study the composition of a biofilm beyond the scope of experimentation by combining different CLSM image stacks of the same sample, and to use the standard techniques for the visualization of flow simulations including streamlines, streaklines, particle tracing, isosurfaces, and so forth (Fig. 3).

The efficient exchange of information—e.g. from CLSM via image analysis tools to a simulation program and further on to some visualization tool—requires a direct and highly streamlined data handling process. This underlines the need for uniform data structures like octrees [70] and efficient (automated) interfaces.

6.4. Interfaces

Direct or automated links of a simulation program to other tools are increasingly important. The large data stacks from CLSM and image processing or from a geometry generation program have to be transformed into a corresponding computational geometry. After the simulation, the results have to be passed over to a visualization tool in order to allow interpretation and validation. Furthermore, there is an increasing need for an efficient interplay of different simulation types, for example when a reactor simulation and a microscopic detail simulation or a flow field and a growth simulation need to be coupled. While such interfaces are starting to be used in research applications, their widespread application is still limited due to the lack of standardized and simple transfer protocols.

7. Matching the potentials of science and engineering

Generally speaking, science is focused on the proliferation of knowledge. Engineers apply knowledge accumulated in science to create new technology and to solve practical problems. Engineers depend on advances gained in science and scientists, to a large extent, depend on engineers to provide them with questions, observations and technological insights to achieve scientific advances. The borders between science and engineering, in particular between microbiology, mathematical modeling, numerical simulation and environmental technology have become somewhat fuzzy during the past years. Microbiologists, under the impression of results of gene probe-based population studies, started to advise process engineers in charge of wastewater treatment system design and operation. Because of insufficient information regarding the complexities of municipal and industrial wastewater management systems, but also of the microbial population dynamics and their influence on treatment processes, the advice provided by microbiologists was not always sustainable or understood properly. In response, engineers argued that the colorful pictures coming out of the laboratories of microbiologists have no practical value, and knowing the name of a "bug" and its position in the phylogenetic tree is of no practical relevance as long as it remains unknown which operational factors are responsible for the presence or absence of a specific bacterial strain, and which role the strain plays within the technical system.

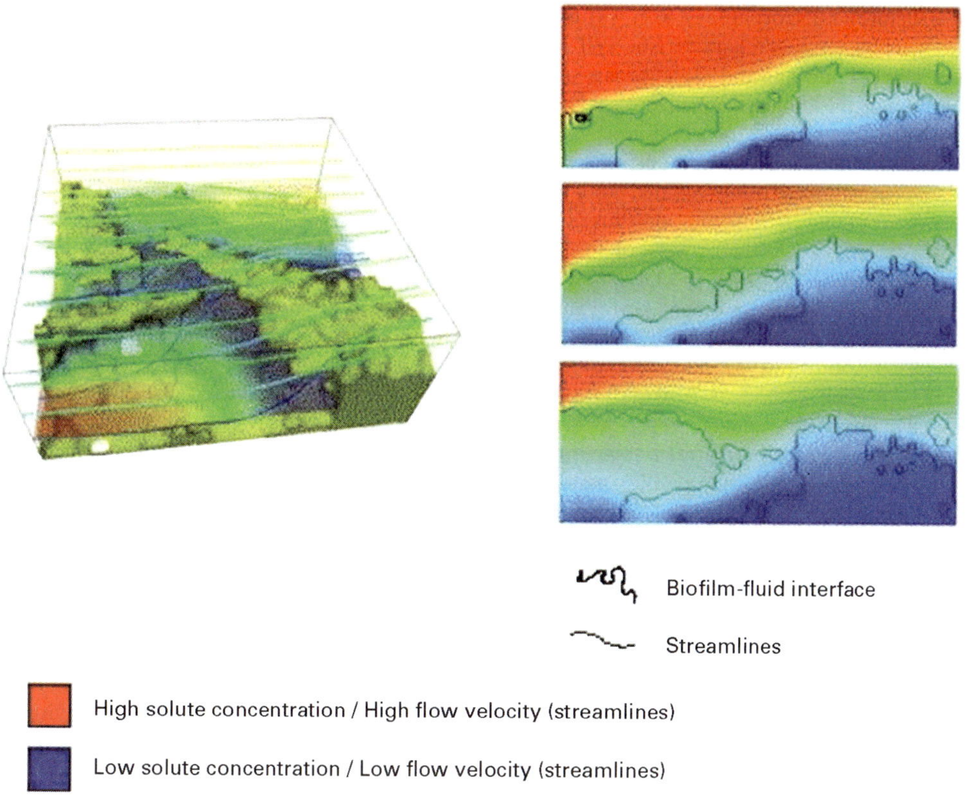

Fig. 3. Numerical simulation of fluid flow, nutrient transport, and biomass growth in defined microscopic biofilms (left: streamlines and nutrient concentration profiles in a *Sphingomonas* sp. strain LB 126 monoculture (3D); right: 2D slices of the growing biofilm with streamlines and nutrient concentrations in the bulk fluid and in the biofilm itself).

In the following, we want to remind the reader that microscopic analysis as well as mathematical modeling has been successfully used for decades to better understand and improve activated sludge and biofilm reactor systems. The only difference is that the methods available nowadays are far more sophisticated and complex than the relatively simple methods used in the past. On the other hand, the treatment processes in use these days, and even more so novel technologies being developed, are also much more complex and demanding than the simple activated sludge tanks used in the past.

The first message is: In order to cope with the rapidly expanding knowledge base in science the engineer must expand his or her scientific knowledge as well. Likewise, the scientists need to be much more aware of the practical aspects of the treatment systems, need to know the principles of modern processes and how they influence the results of the microbial samples taken. The second message is that observations made at the microscale level can only help solve problems on the macroscale when microscale results are correlated with parameter values characteristic for the system as a whole. It is in fact irrelevant to know which types of bacteria are present in a certain activated sludge floc unless the appearance and abundance of certain species or groups of species correlates with the process conditions under which the microbial community has formed, and with the performance data obtained at the reactor in question. In addition, information about microbial systems must be correlated with the performance data of the mixed culture. Only the results of those correlation exercises deliver valuable information needed to draw conclusions relevant to the technical system, which has to be designed, built, optimized, and operated under the time variable conditions typical for wastewater treatment systems in general, and for the very application site in particular.

From the beginning of biological wastewater treatment microscopic investigations have been carried out to gain information about the relationship between the predominant milieu, the substrate characteristics and the composition of the microbial community (e.g. [71–73]). These investigations were—due to limitations of the classic light microscopy—restricted to organisms of conspicuous structure and morphology (e.g., protozoa, metazoa, but also bacteria of characteristic cell morphology, movement, budding, and stalk or aggregate formation). Studying the biological properties of these organisms makes it easy to define conditions, which promote their growth. Physiological properties have in most cases been investigated using isolated and cultivated organisms. Applying the information thus gained to biological wastewater led to the concept of so-called indicator organisms, which 'indicate' special milieu and/or substrate conditions during treatment performance. These indicators have for many years been a useful tool to assess operation conditions by microscopic sludge investigation [74–76]. The method is useful in those cases where simple reactors are applied for wastewater treatment, i.e. where the main treatment objective is the elimination of carbonaceous compounds generally carried out in a single, mostly highly loaded reactor. In such a reactor the concentration of oxygen, carbon compounds (COD or BOD) and biomass content (assumed to be correlated with VSS) can be reliably measured and predicted, and the food to microorganism (F/M) ratio or sludge age can be quickly and redundantly assessed.

However, today's treatment standards are much higher and include biological nitrogen and phosphorus removal. This requires a much more complex operation regime with anoxic and anaerobic zones and an operation at very low F/M ratios imposing various stress situations such as oxygen or substrate deficiency on the organisms. By introducing various recycles such as returning sludge from the secondary clarifier to the anaerobic or anoxic zone, recycling nitrate-loaded mixed liquor to the denitrification zone or returning denitrified sludge to the anaerobic zone, various sludge fractions are mixed forcing us to assess not a one-reactor environment but a mean value over various reactors with quite different environmental conditions. Importantly, it has to be considered that in each of the different compartments in a modern BNR plant, only a particular fraction of the biomass will be active, but all of the biomass is exposed to all conditions repeatedly. Nowadays, operation assessment via microscopic sludge investigation affords sophisticated methods of observation and interpretation.

As an example of how to find a way out of the dead end of light microscopy and benefit from the promising modern analytical methods we might focus on the history of filamentous bacteria investigation as a tool for troubleshooting sludge separation problems such as sludge bulking and scum formation. Sludge bulking, being characterized by a voluminous activated sludge with a high sludge volume index over $150\,ml\,g^{-1}$, and sludge scum formation, being characterized by floating sludge fractions, might end up in sludge losses into the secondary effluent and concomitant deteriorated purification efficiency. These phenomena in general are associated with high numbers of filamentous bacteria [77], which for a long time have been summarized as the 'Abwasserpilz' ('wastewater fungus') *Sphaerotilus natans*. There are, however, many different filamentous microorganisms involved in activated sludge bulking. Eikelboom [78] empirically sorted these bacteria. In municipal wastewater he found around 30 types based on simple morphological features such as cell size and shape, filament length, width and so forth, and created a dichotomous identification key which still forms the basis for assessing sludge separation problems. Many of these filamentous bacteria are, however, still very little characterized and have to be addressed by type numbers [74]. Certain counter-measures have been derived from autecological (that is, studies on isolated microorganisms) data from earlier studies [75,76,79,80].

The large number of filaments being encountered during bulking and scum events led to the assumption that these bacteria are the very culprit. However, the strict differentiation between 'filamentous bacteria' and 'non-filamentous bacteria' (the so-called floc formers) is not permissible from a microbiological point of view, because it is generally accepted that morphology is strongly dependent on the specific environmental conditions prevailing in the reactor [81–83]. Rhodococci, as an example, break into single cells during their stationary phase [84,85] which is a probable growth phase in sludge blankets with a very high sludge age. Grazing, which plays an important role in particular in low F/M systems supporting high numbers of protozoa and metazoa, is known as well to influence the formation of filaments by affecting bacteria growth rates [86,87]. Thus in a low F/M environment typical of current wastewater treatment systems one must not conclude a filamentous organism to be successfully eliminated from the system for the lack of its detection by phase contrast microscopy.

Nowadays, FISH can visualize even inconspicuous single cells. It can help interpret the fate and behavior of 'bulking and scum bacteria' even in those cases where they are actually not growing as filaments [16]. Moreover, taxonomic characterization will benefit from bacterial group-, genus-, or species-specific probes. Eikelboom's type numbers will be replaced and the organisms be assigned their proper phylogenetic affiliation in the near future. Similar "morphotypes" can then be correctly addressed, as it is well known that organisms with a similar morphology might by far not embody a single genetic entity [88]. Some of the aforementioned modern analytical techniques such as microsensors and microautoradiography have yielded promising insights into in situ substrate availability [89] and uptake rates for 'bulking and scum bacteria' in the aerobic [90,91] and in the anoxic and anaerobic environment [92]. However, despite these advances in the microbial identification of filamentous organisms, the understanding of the operational conditions causing them to grow as filaments are still very limited. Since the organisms seem to be present in most sludges, what makes them suddenly grow into large, filamentous colonies is becoming the key question. If more on this was known, measures to control them could be developed from a fundamental understanding of their metabolic and growth characteristics. This would be a major step forward compared to the purely empirical "rules" available now, which have been shown to be not very effective in many cases. One essential element in the development of this knowledge is the detailed investigation and close correlation of both operational and microbial conditions.

The sum of all these new approaches will allow in the near future to predict, assess, model, and finally overcome wastewater treatment problems based on improved knowledge compared to the information available at this time. A database has to be built up which contains information about

- the conditions under which the microorganisms detected in the activated sludge floc or the biofilm have grown (e.g., type of wastewater, mass loading, electron acceptor availability, hydrodynamic conditions, retention time distribution function in the reactor),
- the metabolic activity of the microorganisms in terms of kinetic and stoichiometric parameter values, and
- the detection methods applied including methods of sampling, fixation and storage time.

The appended questionnaire is intended to provide such a record when sampling wastewater treatment plants. The recorded data together with analytical results of the sample (microscopic observations, metabolic rate data, concentration profiles and so forth) should be fed into an internet-based data acquisition system and evaluated using advanced statistical methods, mathematical modeling, and numerical simulation. A sophisticated evaluation is necessary to be able to transform the results obtained at the microscale to the macroscale level. The results should be made available to the scientific and engineering community as a source of information, and as a means to improve design and performance of wastewater treatment plants.

8. Conclusions

This paper has been initiated by the observation of an increasing gap of understanding and communication between the scientific and technical/engineering communities in the environmental field. However, close interactions and knowledge transfer is essential and often highly beneficial for the development and operation of practical, full-scale processes. The rapid developments and associated increase in complexity in some of the specialist methods, e.g., microbiology identification or numerical simulations, might be one cause for the lack of communication and interactions. This paper therefore tried to help bridge this gap by explaining the concepts, benefits and limitations of some of these methods. The aim is to increase communication and cooperation between the various specialists. In particular, the following aspects have been identified as important elements in this understanding:

1. The methods that have been recently developed to study structure and function of microbial aggregates including activated sludge flocs and biofilms have greatly increased our knowledge about the microbial systems, which the performance of wastewater treatment plants depends on. However, the relevance and applicability of this knowledge still needs to be developed strongly to ensure that direct practical benefits from the new findings can be gained.
2. The novel molecular microbiology methods are aimed at providing insights into microscopic structures, but most of the information is of a qualitative nature. These methods have to be applied in concert with quantitative methods of modern microscopy and microsensor analysis. This will allow to directly incorporate the information gained into current design and operating guidelines and particularly to use them for model development and validation. In addition, more specific data are required describing the metabolic potential of the microorganisms on the species level as well as on the community level.
3. Transformation of the results obtained on the microscale requires a thorough correlation with the data describing the reactor systems from which samples were taken. Engineers and microbiologists must focus their efforts and work together to bring their combined insights to bear on wastewater treatment design and operations. The questionnaire presented in the appendix may assist researchers in acquiring the most important information from the plant or reactor sampled.
4. Mathematical modeling and numerical simulation are excellent tools to link information from microscopic and macroscopic scales, from short and long time-frames, and from fundamental and practical knowledge. Therefore, they can provide the knowledge management tool needed to effectively incorporate novel findings into the practical process of designing and operating advanced reactor systems for biological wastewater treatment.
5. It does matter which organisms perform important steps. Their specific physiological requirements must

be recognized and met during operation. Failure to do so can lead to inefficient performance, for example, of nitrification and biological P removal steps.

6. Certain modern analytical techniques are amenable to routine applications. FISH is recommended for the monitoring of plant operation by testing laboratories and plant operators. Training in the use of fluorescence microscopy, decisions regarding the choice of gene probes, and interpretations of FISH results should involve the services of an experienced consultant.
7. The greatest challenge in applied microbiology of wastewater treatment is likely the structure-function analysis. The identification and in situ location of microbial communities must be coupled with measurements of activity and local conditions to obtain usable information about process operation.

Acknowledgements

We thank E. Müller for critically reading the manuscript and S. Hilber for editorial assistance. This work was supported by the Research Center for Aerobic Biological Wastewater Treatment (SFB 411) in Munich, Germany through funding from the German Research Foundation (DFG).

Appendix

Biography

of a microbial sample

Sample: ______________________ from ______________________

sample name, description — date/time of sample collection

____

official in charge (scientist, engineer...)

Results published?

☐ no

☐ yes ____

published in______________________

Note:

This ***Biography*** contains questions to the origin and history of a biological sample in respect of milieu, technical apparatus and preparatory and analytical methods. Please, answer every sample-related question, if possible.

Part 1—Milieu

	Treated Wastewater	Influent and Milieu of the Reactor Examined **(average over day)**		
☐	municipal wastewater	COD	[mg/L]	________
☐	municipal wastewater with significant industrial proportions specify	BOD_5	[mg/L]	________
☐	industrial wastewater specify	NH_4-N	[mg/L]	________
☐	other ________	NO_3-N	[mg/L]	________

PO_4-P	[mg/L]	______________
susp. solids	[mg/L]	______________
temperature	[°C]	______________
pH value	[–]	______________
O_2	[mg/L]	______________
conductivity	[μS/cm]	______________
alkalinity	[mmol/L]	______________

Part 2—System

Flow-Schematic with Sampling Location (sketch or in words)

Total System

Scale

☐ **lab scale**

☐ **pilot scale**

☐ **full scale**

Treatment System

☐ **single step treatment plant**

☐ **multi stage treatment plant**

☐ ______________

Design / Inflow

capacity (popul. equ.) [cp] ______________

average inflow

(dry weather flow) [ ] ______________

Operational State

☐ **start-up operation**

☐ **stable operation without breakdowns**

☐ **alternating operation (campaign...)**

☐ **recent loading peaks**

specify ______________

Reactor Examined

Technique

☐ **activated sludge system**

☐ **biofilm reactor**

☐ ______________

Operation Mode

☐ **continuous flow**

☐ **cyclically operated (SBR, SBBR…)**

☐ ______________

Inflow incl. recirculation (day of sample collection)

Q_{in} **[ ]** ______________

Treatment Objectives

(multiple selection possible)

☐ **COD removal**

☐ **nitrification**

☐ **denitrification**

☐ **enhanced biol. phosphorus removal**

☐ **P-precipitation with** ______________

☐ **other** ______________

Questions to Biofilm Reactors (reactor examined)

Type of Reactor

☐ Trickling Filter

☐ Rotating Biological Contactor

☐ Submerged Fixed-Bed Reactor

Support Media

☐ mineral media

☐ plastic media

☐ other ______________

	□ packed bed	size of media	[mm]	______
	□ structured media	voids fraction	[%]	______
□ Fluidized Bed Reactor		spec. surface area	[m^2/m^3]	______
□ Hybrid-System (activated sludge-biofilm)				
	□ fixed support media	height of fixed bed	[m]	______
	□ floating support media	cross section area	[m^2]	______

Washing Strategy		**Sampling Location**		
□ continuous washing		vertical distance from influent	[m]	______
□ intermittent washing washing frequency	[1/d] ______	horizontal distance from center line	[m]	______
operation time since last washing	[h] ______	biofilm thickness	[μm]	______

Operational Characteristics

Continuous-Flow System

filter velocity (incl. recirculation)	**[m/s]**	______
hydraulic retention time	**[h]**	______

Multiple Zone Reactors

aerobic zone	**[h]**	______
anoxic zone	**[h]**	______
anaerobic zone	**[h]**	______

Cyclically Operated System (e.g., SBBR)

vol. exchange rate [%]		______
filling strategy	□	**quick fill**
	□	**slow fill**
Recirculation Period		
filter velocity	**[m/s]**	______

Times of Phases

fill	**[h]**	______
aerobic phase	**[h]**	______
anoxic phase	**[h]**	______
anaerobic phase	**[h]**	______
draw	**[h]**	______

Questions to Activated Sludge Systems (reactor examined)

Actual Operation Characteristics

sludge age	[d]	______
aerobic sludge age	[d]	______
biomass concentration (MLSS aeration tank)	[g SS/L]	______
sludge volume index SVI	[ml/g]	______

Aeration System

□ surface aeration

□ diffused air aeration

□ continuous aeration

□ intermittent aeration

Operational Characteristics

Continuous Flow System		**Sequencing Batch Reactor**		
☐	completely mixed reactor	volume	[m^3]	________
☐	reactor cascade	exchange rate	[%]	________
☐	plug flow reactor	(mean)		
Hydraulic Retention Times		**Times of Phases**		
aerobic	[h]	fill	[h]	________
anoxic	[h]	aerobic	[h]	________
anaerobic	[h]	anoxic	[h]	________
settlement tank	[h]	anaerobic	[h]	________
		settling phase	[h]	________
		draw	[h]	________

Comments

____

____

____

____

____

____

____

Part 3—Preparatory and Analytical Methods

Microscopic Analysis

Time passed between sampling and microscopic analysis

☐ Sample kept at 4–10°C during transport and storage?

☐ Sample homogenized and/or centrifuged? (specify)

Conventional Light Microscopy

☐ Structure of flocs recorded

☐ Filaments identified by guide (Eikelboom, Jenkins)

☐ Protozoa identified

☐ Metazoa identified

☐ Simple staining procedures applied
specify____

☐ Any unusual observations
specify____

Fluorescence Microscopy

☐ Total bacterial cell count (e.g. DAPI staining)
specify____

☐ Viable cell count (e.g. CTC staining)
specify____

☐ Other
specify____

Fluorescent In Situ Hybridization (FISH)

Sample Fixation

☐ 4% PFA

☐ Ethanol

☐ Additional treatment (specify)

Sample was

☐ Homogenized

☐ Not homogenized

☐ Either of the above plus tested the same day for total bacterial cell count (ethanol-fixed cells)

Instrumentation used

☐ Fluorescence microscope

☐ Confocal laser scanning microscope

☐ Either of above plus image analysis

Cell Counts

☐ Manual

☐ Automatic

Fluorescent labels used

____

____

Has the reactor been analyzed by FISH or related methods before?

☐ No

☐ Yes

Was a 16S/23S rRNA gene bank prepared?

☐ No

☐ Yes

Have reactor-specific gene probes for FISH been developed?

☐ No

☐ Yes

Gene probes used for FISH

Comments

This form sheet is available for downloading as a WORD file from www.wga.bv.tum.de. Please click on database. The completed questionnaire may be sent via e-mail. All submitted data will be incorporated in a database with open access for downloading.

References

[1] Amann RI, Ludwig W, Schleifer KH. Phylogenetic identification and in situ detection of individual microbial cells without cultivaton. Microbiol Rev 1995;59:143–69.

[2] Purkhold U, Pommerening-Röser A, Juretschko S, Schmid MC, Koops HP, Wagner M. Phylogeny of all recognized species of ammonia-oxidizers based on comparative 16S rRNA and *amoA* sequence analysis: implications for molecular diversity surveys. Appl Environ Microbiol 2000;66:5368–82.

[3] Nielsen AT, Liu WT, Filipe C, Grady jr L, Molin S, Stahl DA. Identification of a novel group of bacteria in sludge from a deteriorated biological phosphorus removal reactor. Appl Environ Microbiol 1999;65:1251–8.

[4] Delbes C, Moletta R, Godon JJ. Monitoring of activity dynamics of an anaerobic digester bacterial community using 16S rRNA polymerase chain reaction-single-strand conformation polymorphism analysis. Environ Microbiol 2000;2:506–15.

[5] Liu WT, Marsh TL, Forney LJ. Determination of the microbial diversity of anaerobic-aerobic activated sludge by a novel molecular biological technique. Water Sci Technol 1998;37:417–22.

[6] de los Reyes FL, Ritter W, Raskin L. Group-specific small-subunit rRNA hybridization probes to characterize filamentous foaming in activated sludge systems. Appl Environ Microbiol 1997;63:1107–17.

[7] Schuppler M, Wagner M, Schön G, Göbel UB. In situ identification of nocardioform actinomycetes in activated sludge using fluorescent rRNA-targeted oligonucleotide probes. Microbiology 1998;144:249–59.

[8] Kuehn M, Hausner M, Bungartz HJ, Wagner M, Wilderer PA, Wuertz S. Automated confocal laser scanning microscopy and semiautomated image processing for analysis of biofilms. Appl Environ Microbiol 1998;64:4115–27.

[9] Wagner M, Hutzler P, Amann R. 3-D analysis of complex microbial communities by combining confocal laser scanning microscopy and fluorescence in situ hybridization (FISH). In: Wilkinson MH, Schut F, editors. Digital Image Analysis of Microbes: Imaging, Morphometry, Fluorometry and Motility Techniques and Application. Chichester: John Wiley & Sons Ltd., 1998. p. 467–86.

[10] Lee N, Nielsen PH, Andreasen KH, Juretschko S, Nielsen JL, Schleifer KH, Wagner M. Combination of fluorescent in situ hybridization and microautoradiography—a new tool for structure-function analyses in microbial ecology. Appl Environ Microbiol 1999;65:1289–97.

[11] Schramm A, Larsen LH, Revsbech NP, Ramsing NB, Amann R, Schleifer KH. Structure and function of a nitrifying biofilm as determined by in situ hybridization and the use of microelectrodes. Appl Environ Microbiol 1996;62:4641–7.

[12] Schramm A, de Beer D, Wagner M, Amann R. Identification and activities in situ of *Nitrosospira* and *Nitrospira* spp. as dominant populations in a nitrifying fluidized bed reactor. Appl Environ Microbiol 1998;64:3480–5.

[13] Saiki RK, Gelfand DH, Stoffel S, Scharf SJ, Higuchi R, Horn GT, Mullis KB, Ehrlich HA. Primer-directed enzymatic amplification of DNA with a thermostable DNA polymerase. Science 1988;239:487–91.

[14] Wagner M, Amann R, Kämpfer P, Assmus B, Hartmann A, Hutzler P, Springer N, Schleifer KH. Identification and

in situ detection of gram-negative filamentous bacteria in activated sludge. System App Microbiol 1994;17: 405–17.

[15] Kanagawa T, Kamagata Y, Aruga S, Kohno T, Horn M, Wagner M. Phylogenetic analysis of and oligonucleotide probe development for Eikelboom type 021N filamentous bacteria isolated from bulking activated Sludge. Appl Environ Microbiol 2000;66:5043–52.

[16] Lemmer H, Lind G, Müller E, Schade M, Ziegelmayer B. Scum in activated sludge plants: impact of non-filamentous and filamentous bacteria. Acta Hydrochim Hydrobiol 2000;28:34–40.

[17] Juretschko S, Timmermann G, Schmid M, Schleifer KH, Pommerening-Röser A, Koops HP, Wagner M. Combined molecular and conventional analyses of nitrifying bacterium diversity in activated sludge: *Nitrosococcus mobilis* and *Nitrospira*-like bacteria as dominant populations. Appl Environ Microbiol 1998;64:3042–51.

[18] Rittmann B, Laspidou CS, Flax J, Stahl DA, Urbain V, Harduin H, van der Waarde JJ, Geurkink B, Henssen MJ, Brouwer H, Klapwijk A, Wetterauw M. Molecular and modeling analyses of the structure and function of nitrifying activated sludge. Water Sci Technol 1999; 39:51–9.

[19] Strous M, Kuenen JG, Jetten MSM. Key physiology of anaerobic ammonium oxidation. Appl Environ Microbiol 1999;65:3248–50.

[20] Schmid M, Twachtmann U, Klein M, Strous M, Juretschko S, Jetten M, Metzger JW, Schleifer KH, Wagner M. Molecular evidence for genus level diversity of bacteria capable to catalyze anaerobic ammonium-oxidation. Syst Appl Microbiol 2000;23:93–106.

[21] Crocetti GR, Hugenholtz P, Bond PL, Schuler A, Keller J, Jenkins D, Blackall LL. Identification of polyphosphate accumulating organisms and the design of 16S rRNA-directed probes for their detection and quantitation. Appl Environ Microbiol 2000;66:1175–82.

[22] Burrell P, Keller J, Blackall LL. Microbiology of a nitrite oxidizing bioreactor. Appl Environ Microbiol 1998;64:1878–83.

[23] Felske A, Engelen B, Nübel U, Backhaus H. Direct ribosomal isolation from soil to extract bacterial rRNA for community analysis. Appl Environ Microbiol 1996;62:4162–7.

[24] Muyzer G, Smalla K. Application of denaturing gradient gel electrophoresis (DGGE) and temperature gradient gel electrophoresis (TGGE) in microbial ecology. Antonie Van Leewenhoek 1998;73:127–41.

[25] Murray AE, Hollibaugh JT, Orrego C. Phylogenetic composition of bacterioplankton from two California estuaries compared by denaturing gradient electrophoresis of 16S rDNA fragments. Appl Environ Microbiol 1996;62:2676–80.

[26] Felske A, Engelen B, Nübel U, Backhaus H. Direct ribosome isolation from soil to extract bacterial rRNA for community analysis. Appl Environ Microbiol 1996; 62:4162–7.

[27] Donner G, Schwarz K, Hoppe HG, Muyzer G. Profiling the sucession of bacterial populations in pelagic chemoclines. Arch Hydrobiol Special Issues Advance Limnol 1996;48:7–14.

[28] Wawer C, Jetten M, Muyzer G. Genetic diversity and expression of the (NiFe) hydrogenase large subunit gene of *Desulfovibrio* spp. in environmental samples. Appl Environ Microbiol 1997;63:4360–9.

[29] Ward DM, Santegoeds CM, Nord SC, Ramsing NB, Ferris MJ, Bateson MM. Biodiversity within hot spring microbial mat communities: molecular monitoring of enrichment cultures. Antonie Van Leeuwenhoek 1996;71: 143–50.

[30] Surmann SB, Walker JT, Goddard DT, Morton LHG, Keevil CW, Weaver W, Skinner A, Hanson K, Caldwell DKJ. Comparison of microscope techniques for the examination of biofilms. J Microbiol Meth 1996;25: 57–70.

[31] Viles CL, Sieracki ME. Measurement of marine picoplankton cell size by using a cooled, charge-coupled device camera with image-analyzed fluorescence microscopy. Appl Environ Microbiol 1992;58: 584–92.

[32] An YH, Friedman RJ, Draughn RA, Smith EA, Nicholson JH, John JF. Rapid quantification of staphylococci adhered to titanium surfaces using image analyzed epifluorescence microscopy. J Microbiol Meth 1995; 24:29–40.

[33] Bloem J, Veninga M, Shepherd J. Fully automatic determination of soil bacterium numbers, cell volumes, and frequencies of dividing cells by confocal laser scanning microscopy and image analysis. Appl Environ Microbiol 1995;61:926–36.

[34] Bouchez T, Patureau D, Dabert P, Juretschko S, Doré J, Delgenès P, Moletta R, Wagner M. Ecological study of a bioaugmentation failure. Environ Microbiol 2000;2: 179–90.

[35] Korber DR, Lawrence JR, Hendry MJ, Caldwell DE. Programs for determining statiscally representative areas of microbial biofilms. Binary 1992;4:204–10.

[36] Heydorn A, Nielsen AT, Hentzer M, Sternberg C, Givskov M, Ersboll BK, Molin S. Quantification of biofilm structures by the novel computer program Comstat. Microbiology 2000;146:2395–407.

[37] Xavier JB, Schnell A, Wuertz S, Palmer R, White DC, Almeida JS. Objective threshold selection procedure (OTS) for segmentation of scanning laser confocal microscope images. J Microbiol Meth, in press.

[38] Lewandowski Z, Webb D, Hamilton M, Harkin G. Quantifying biofilm structure. Water Sci Technol 1999;39:71–6.

[39] Xavier JB, Mahlo R, Reis AM, Almeida JS. Description of biofilm formation by determination of developmental axis. Water Sci Technol 2000;41:121–7.

[40] Daims H, Nielsen PH, Nielsen JL, Juretschko S, Wagner M. Novel *Nitrospira*-like bacteria as dominant nitrite-oxidizers in biofilms from wastewater treatment plants: diversity and in situ physiology. Water Sci Technol 2000;41:85–90.

[41] de Beer D. Use of micro-electrodes to measure in situ microbial activities in biofilms, sediments and microbial mats. In: Akkermans AOL, van Elsas JO, de Bruijn FD, editors. Molecular Microbial Ecology Manual, 8, 1.3, Vol. 8. The Netherlands: Kluwer Academic Publishers, 1999. p. 1–23.

[42] Revsbech NP, Christensen PB, Nielsen LP, Sørensen SJ. A combined oxygen and nitrous oxide microsensor for denitrification studies. Appl Environ Microbiol 1988;45:2245–9.

[43] Jeroschewski P, Steuckart C, Kühl M. An amperometric microsensor for the determination of H_2S in aquatic environments. Anal Chem 1996;68:4351–7.

[44] de Beer D, Sweerts JP. Measurement of nitrate gradients with an ion-selective microelectrode. Anal Chem 1989;219:35–356.

[45] Revsbech NP. Diffusion characteristics of microbial communities determined by use of oxygen microsensors. J Microbiol Meth 1989;9:111–22.

[46] de Beer D, van den Heuvel JC. Response of ammonium-selective microelectrodes based on the neutral carrier nonactin. Talanta 1988;35:728–30.

[47] Hinke J. Glass microelectrodes for the study of binding and compartmentalisation of intracellular ions. In: Lavallee M, Schanne OF, Herbert NC, editors. Glass Microelectrodes. New York: Wiley and Sons, 1969. p. 349–75.

[48] van den Heuvel JC, de Beer D, Cronenberg CCH. Microelectrodes: A versatile tool in biofilm research. In: Biofilms—Science and Technology. Dordrecht: Kluwer Academic Publishers, 1992. p. 631–44.

[49] Amann D, Bührer T, Schefer U, Müller M, Simon W. Intracellular neutral carrier based Ca^{2+}microelectrode with sub-nanomolar detection limit. Pflügler's Arch 1987;409:223–38.

[50] Khuri RN, Bogharian KK, Agulian SK. Intracellular bicarbonates in single skeletal muscle fibers. Pflüglers Arch 1974;349:285–94.

[51] Schulthess P, Shijo Y, Pham HV, Pretsch E, Ammann D, Simon W. A hydrogen ion-selective liquid-membrane electrode based on tri-n-dodecylamine as neutral carrier. Anal Chim Acta 1981;131:111–6.

[52] Klimant I, Kühl M, Glud RN, Holst G. Optical measurements of oxygen and temperature in microscale: strategies and biological applications. In: Sensors and Actuators B, 1997, pp. 38–39, 29–37.

[53] Lawrence JR, Wolfaardt GM, Korber DR. Determination of diffusion coefficients in biofilm by confocal laser scanning microscopy. Appl Environ Microbiol 1994;60:1166–73.

[54] Santegoeds C, Schramm A, de Beer D. Microsensors as a tool to determine chemical microgradients and bacterial activity in wastewater biofilms and flocs. Biodegradation 1998;9:159–67.

[55] de Beer D, Stoodley PD, Lewandowsky Z. Effects of biofilm structures on oxygen distribution and mass transfer. Biotechnol Bioeng 1994;43:1131–8.

[56] Hermanowicz SW, Schindler U, Wilderer PA. Fractal structure of biofilms—new tools for investigation of morphology. Water Sci Technol 1995;32:99–105.

[57] Zhang TC, Fu Y, Bishop PL. Competition in biofilm. Water Sci Technol 1994;29:263–70.

[58] Morgenroth E, Eberl H, van Loosdrecht MC. Evaluating 3-D and 1-D mathematical models for mass transport in hererogenous biofilms. Water Sci Technol 2000;41: 347–56.

[59] Bungartz HJ, Kuehn M, Mehl M, Hausner M, Wuertz S. Fluid flow, transport and biomass growth in defined biofilms: experiments and numerical simulations on a microscale. Water Sci Technol 2000;41:331–8.

[60] Picioreanu C, van Loosdrecht MC, Heijnen JJ. Mathematical modeling of biofilm structure with a hybrid differential-discrete cellular automation approach. Biotechnol Bioeng 1998;58:101–16.

[61] Reichert P. A tool for simulation and data analysis of aquatic systems. Water Sci Technol 1994;30:21–30.

[62] Horn H, Wulkow M. RIOVAL: A management tool for water quality modelling in small streams. http://www.wst.shuttle.de/cit/p1rioval.pdf, 1999.

[63] de Beer D, Stoodley PD, Lewandowsky Z. Liquid flow and mass transport in heterogeneous biofilms. Water Res 1996;30:2761–5.

[64] Wanner O, Reichert P. Mathematical-modeling of mixed-culture biofilms. Biotechnol Bioeng 1996;49:172–84.

[65] Picioreanu C, van Loosdrecht MC, Heijnen JJ. Discrete-differential modelling of biofilm structure. Water Sci Technol 1999;39:115–22.

[66] Tartar L. Incompressible Fluid Flow in a Porous Medium—Convergence of the Homogenization Process. Tartar, L., Springer, Lecture Notes in Physics 129 ed., 1980. p. 368–77.

[67] Krafczyk M, Lehmann P, Filippova O, Hänel D, Lantermann U. Lattice Boltzmann simulations of complex multiphase flows. In: Wendland W, editor. Multifield Problems in Solid and Fluid Mechanics. Berlin: Springer Proceedings, 1999.

[68] Babovski H. Die Boltzmann-Gleichung In: Modellbildung–Numerik–Anwendungen. Teubner, B.G: Stuttgart & Leipzig, 1998.

[69] Griebel M, Dornseifer T, Neunhoeffer T. Numerische Simulation in der Strömungsmechanik: eine praxisorientierte Einführung. Vieweg Lehrbuch Scientific Computing. Braunschweig, 1995.

[70] Frank A. Organizational principles for the integration of geometric modeling, numerical simulation and visualization. PhD Thesis, TU München, 2000.

[71] Liebmann H. Handbuch der Frischwasser- und Abwasser-Biologie, Oldenbourg Verlag München, 1962.

[72] Bick H. Ciliated Protozoa. World Health Organisation 1972;1–198.

[73] Curds CR, Hawkes HA. Ecological aspects of used-water treatment. London: Academic Press, 1975.

[74] Eikelboom DH, van Buijsen HJ. Microscopic sludge investigation manual. IMG-TNO report A94a. The Netherlands: Delft, 1981.

[75] Lemmer H. Fadenförmige Mikroorganismen aus belebtem Schlamm. Vorkommen-Biologie-Bekämpfung. ATV-Dokumentation Und Schriftenreihe Aus Wissenschaft Und Praxis, 1992;30.

[76] Jenkins D, Richard MG, Daigger GT. Manual on the causes and control of activated sludge bulking and foaming, 2nd ed. Boca Raton, USA: Lewis Publishers, 1993.

[77] Smit J. Über die Ursachen des Aufblähens von Belebtschlamm. Arch Mikrobiol 1934;5:550–60.

[78] Eikelboom DH. Filamentous organisms observed in activated sludge. Water Res 1975;9:365–88.

[79] Eikelboom DH. Microscopic sludge investigation manual (incl. CD-ROM). IWA publishing, 2000.

[80] Lemmer H, Lind G. Blähschlamm, Schaum, Schwimmschlamm. Mikrobiologie und Gegenmaßnahmen. Hirthammer, München, 2000.

[81] Buali AM, Horan NJ. Variable morphology in certain filamentous bacteria and the implications of this for theories of activated sludge bulking. Environ Technol Lett 1989;10:941–50.

[82] Foot RJ, Kocianova E, Forster CF. Variable morphology of *Microthrix parvicella* in activated sludge systems. Water Res 1992;26:875–80.

[83] Soddell JA, Seviour RJ. Incidence and morphological variability of *Nocardia pinensis* in Australian activated sludge plants. Water Res 1994;28:2343–51.

[84] Davenport RJ, Elliott JN, Curtis TP, Upton J. In situ detection of rhodococci associated with activated sludge foams. Antonie Van Leeuwenhoek 1998;74:41–8.

[85] Lemmer H, Lind G, Schade M, Ziegelmayer B. Autecology of scum producing bacteria. Water Sci Technol 1998;37:527–30.

[86] Hahn MW, Moore ER, Höfle MG. Bacterial filament formation, a defense mechanism against flagellate grazing, is growth rate controlled in bacteria of different phyla. Appl Environ Microbiol 1999;65:25–35.

[87] Jürgens K, Pernthaler J, Schalla S, Amann RI. Morphological and compositional changes in a planktonic bacterial community in response to enhanced protozoan grazing. Appl Environ Microbiol 1999;65:1241–50.

[88] Bradford D, Hugenholtz P, Seviour EM, Cunningham MA, Stratton H, Seviour RJ, Blackall LL. 16S rRNA analysis of isolates obtained from gram-negative, filamentous bacteria micromanipulated from activated sludge. Syst Appl Microbiol 1996;19:334–43.

[89] Schramm A, Santegoeds CM, Nielsen HK, Ploug H, Wagner M, Pribyl M, Wanner J, Amann RI, de Beer D. On the occurrence of anoxic microniches, denitrification, and sulfate reduction in aerated activated sludge. Appl Environ Microbiol 1999;65:4189–96.

[90] Andreasen K, Nielsen PH. Application of microautoradiography to the study of substrate uptake by filamentous microorganisms in activated sludge. Appl Environ Microbiol 1997;63:3662–8.

[91] Andreasen K, Nielsen PH. In situ characterization of substrate uptake by Microthrix parvicella using microautoradiography, Water Sci Technol 1998;4–5

[92] Andreasen K, Nielsen PH. Growth of *Microthrix parvicella* in nutrient removal activated sludge plants: studies of *in situ* physiology. Water Res 1999;34:1559–69.

[93] Hausner M, Hendrickx L, Rodenacker K, Mehl M, Bungartz H-J, Wuertz S. Spatial distribution of conjugative gene transfer in biofilms, submitted for publication.

Glossary

CLSM (confocal laser scanning microscopy): In conventional light microscopy all light passing through the specimen is detected. In CLSM only some of the light is allowed to pass through the confocal pinhole so that light above and below the plane of focus is excluded. The technique is particularly useful in conjunction with fluorescent signals such as those emitted by a fluorescently labeled nucleic acid probe bound to rRNA in a bacterial cell. If the stage on which the objective slide with the specimen is mounted is moved in the third (z) dimension different optical sections of the specimen are obtained. The resulting 2-D images from different focal planes can be stored digitally and subjected to quantitative image analysis. The method allows spatial analysis of 3-D aggregates.

DGGE (denaturing gradient gel electrophoresis): DNA fragments (for example of 16S rRNA genes) that have been amplified by PCR can be separated according to size and sequence content by DGGE. It is one of several nucleic acid fingerprinting methods available to screen wastewater reactors. The patterns obtained are used to assess changes in the microbial population and the amplified DNA bands are available for sequencing to verify the identity of the organism. The method is not quantitative.

DNA (deoxyribonucleic acid): DNA consists of deoxyribonucleotides which are composed of a base, the sugar deoxyribose and a phosphate group. Two strands of DNA are wrapped together forming a double helix. DNA bases are adenine (A), cytosine (C), thymine (T) and guanine (G). The sugar and phosphate groups form the backbone of the molecule supporting the bases which jut out from the chain. A and T form hydrogen bonds as do C and G thus holding together the double helix. DNA carries the information for the synthesis of RNA and proteins in regions called genes.

FISH (fluorescent in situ hybridization): Oligonucleotides consist of 18 (± 3) nucleotides that are labeled with a fluorescent dye. The resulting molecule is small enough to enter bacterial cells as well as cells of other microorganisms and bind to rRNA associated with ribosomes. Since rRNA is single stranded, hybridization of complementary DNA sequences (that is, sequences with the corresponding A for every U or T and G for every C, and vice versa) in intact cells can be performed. By adjusting the hybridization conditions (the stringency) only those cells which contain the complimentary rRNA sequence to the fluorescently labeled nucleic acid probe will bind the probe. Visualization can be achieved by epifluorescence microscopy or CLSM. The method is quantitative.

Gene: See DNA.

Gene library: A mixture of DNA fragments (e.g. obtained after PCR) is preserved by inserting them into a plasmid (a circular piece of DNA that can replicate independently of the chromosome inside a cell) and introducing the plasmid into a bacterial culture of *Escherichia coli*. This step is referred to as cloning. Normally, each plasmid will contain only one DNA fragment and each bacterial cell will take up only one plasmid molecule. Hence the different fragments are maintained in the cell culture which serves as a repository. The individual bacterial colonies which are formed when a portion of the culture is spread out on solid growth medium are due to the division of one original cell. They are clones of this cell and each cell contains an identical cloned DNA fragment inside the plasmid.

Nucleotide: The smallest unit of a nucleic acid consisting of a base, a sugar and a phosphate group.

Nucleic acid probe (gene probe): An oligonucleotide that can hybridize to a complementary nucleic acid sequence (RNA or DNA). A useful way to confirm the presence or absence of certain genes or microorganisms. Gene probes may be used on extracted nucleic acids or on whole cells, for example in FISH of activated sludge.

MAR (microautoradiography): A radioactively labeled substrate is fed to microorganisms to obtain information about utilization rates and substrate specificity. If MAR is combined with FISH the in situ activities of individual bacterial cells can be analyzed. For example, using this approach it is possible to determine whether certain bacteria can function in biological phosphorus removal.

Oligonucleotide: Several nucleotides in the DNA strand joined by phosphodiester bonds.

PCR (polymerase chain reaction): Any DNA fragment can be generated from a template strand and amplified using the enzyme DNA polymerase from a thermophilic bacterium. Two primers are needed to make copies of a given region of DNA located between sequences complementary to the primers. A PCR cycle consists of 3 different steps which occur at different temperatures: denaturing of the target DNA (leading to the formation of single strands), annealing of the primers (binding to the template DNA), and DNA synthesis. Typically this cycle is repeated 20 to 40 times during which time the target DNA is amplified exponentially.

Primer: A single-stranded RNA or DNA that is complementary to a single-stranded DNA template. By binding to the latter it provides a free 3′ hydroxyl end to which the enzyme DNA polymerase can add deoxynucleotides. In this way a new strand of DNA complementary to the template DNA is synthesized.

Ribosome: A macromolecular assembly of rRNAs and proteins which serves as the site of protein synthesis. Bacterial ribosomes contain three types of RNA, namely, 16S, 23S, and 5S.

rDNA: The genes coding for rRNA.

RNA (ribonucleic acid): A single-stranded type of nucleic acid which contains the base uracil (U) instead of thymine (T) found in DNA and the sugar ribose instead of deoxyribose.

mRNA (messenger RNA): Genes are transcribed (or copied) from one strand to RNA. An RNA molecule carrying information for a protein is called mRNA. It has a short half-life in bacterial cells.

rRNA (ribosomal RNA): A class of RNA molecules serving as components of ribosomes. They are functionally conserved and present in all organisms. In bacteria there are 16S and 23S rRNAs. The primary structures of rRNA contain regions of higher and lower evolutionary conservation which lend themselves to the design of complementary nucleic acid probes for whole groups of bacteria in the form of phylogenetic sublineages (resulting in a crude description of the bacterial community in a specific wastewater sample) or individual genera or species. rRNAs are very common and stable.

16S rRNA: See rRNA

23S rRNA: See rRNA

PERGAMON

Water Research 36 (2002) 394–402

WATER RESEARCH

www.elsevier.com/locate/watres

Microbiological aspects of a bioreactor with submerged membranes for aerobic treatment of municipal wastewater

R. Witzig[a], W. Manz[a], S. Rosenberger[b], U. Krüger[b], M. Kraume[b], U. Szewzyk[a,*]

[a] *Department of Microbial Ecology, Technical University of Berlin, Franklinstrasse 29, Sekretariat OE 5, D-10587 Berlin, Germany*

[b] *Department of Chemical Engineering, Technical University of Berlin, Strasse des 17. Juni 136, Sekretariat MA 5-7, D-10623 Berlin, Germany*

Received 20 March 2000; received in revised form 29 December 2000; accepted 1 February 2001

Abstract

An aerobic membrane bioreactor treating municipal wastewater at complete biomass retention was studied in respect of microbiological parameters over a period of 380 days. The results were compared to those obtained from a conventional activated sludge wastewater treatment plant (WWTP) treating the same wastewater. Microscopically, significant changes in the structure of the flocs and of the ratio between free suspended and aggregated cells could be observed. The presence of filamentous bacteria varied from almost not present to very high numbers. With the exception of short periods after changes in operating conditions, protozoa and metazoa were rarely present in the sludge community. The rate of oxygen consumption and the cell detectability by fluorescence in situ hybridization (FISH) with rRNA-targeted oligonucleotide probes were used to assess the physiological state of the bacterial cells. Oxygen consumption rates of sludge samples obtained from both the conventional and membrane filtration plant were determined without and after addition of different energy and carbon sources. In contrast to the conventional activated sludge, a pronounced increase in respiration activity upon the addition of organic substrates could be observed in the membrane filtration sludge. In situ probing with the *Bacteria*-specific probe EUB338 visualized 40–50% of all DAPI-stainable bacteria in the membrane bioreactor, compared to 80% cells detectable by FISH in the conventional activated sludge. These results suggest that bacteria present in the highly concentrated biomass of the membrane reactor used the energy supplied for their maintenance metabolism and were not in a physiological state characteristic for growth. This assumption could explain the zero net biomass production observed in the reactor.

Keywords: Membrane bioreactor; Complete biomass retention; Metabolic activity; Substrate limitation; Maintenance metabolism; FISH

1. Introduction

The treatment of municipal wastewater by conventional techniques, like activated sludge process or trickling filters, is successful in reducing the content of organic carbon by up to 95–98%. If different treatment steps are combined, even removal of nitrogen and phosphorous can efficiently be achieved by means of biological processes. However, there are still further improvements to be realized concerning the degradation of recalcitrant organic compounds and the hygienical quality of the treated water.

It has long been known that the degradation capabilities of activated sludge treatment systems are limited since, for instance, dyes from textile industry are well known to pass treatment plants in such concentrations that receiving water bodies are completely colored.

*Corresponding author. Tel./fax: +49-30-314-73461.
E-mail address: ulrich.szewzyk@tu-berlin.de (U. Szewzyk).

0043-1354/02/$ - see front matter
PII: S0043-1354(01)00221-4

The detection of pharmaceutical residues in surface waters and even in ground water has demonstrated that quite a range of chemical compounds can pass activated sludge systems without degradation, resulting in their release into the environment [1].

Additionally, increasing concern is directed towards the hygienical quality of the treated water. Growing epidemiological evidence for a correlation between the use of surface waters for recreational purposes, especially bathing and diving, and the occurrence of infections of the ears and eyes [2], has posed the question of wastewater treatment plants as sources for the pathogens. There is therefore considerable need for implementation of reactor configurations suitable to reduce the risks caused by the operation of conventional wastewater treatment facilities.

As one of the most innovative approaches to restrain the release of pathogens from wastewater treatment plants, the use of microfiltration membranes has been recommended [3]. These membranes can either be used as a final clarification step after secondary clarifier passage or be integrated directly into the activated sludge tank. The latter application received increasing attention for two reasons during the last years: (i) the number of bacteria can be reduced to values meeting bathing water requirements and (ii) the activated sludge process can be operated in a way which allows to cope with many of the limitations known to be common in conventional activated sludge processes. As a consequence of the complete biomass retention, the mixed liquor suspended solids (MLSS) concentrations can be maintained at values of 20–60 $g\,l^{-1}$; whereas activated sludge systems are typically operated at MLSS concentrations of 3–4 $g\,l^{-1}$, due to low settling velocities at high sludge concentrations. Additionally, two further drawbacks of activated sludge processes, the wash out of slow growing bacteria and the bulking phenomenon due to massive development of filamentous bacteria, do not longer impair the process control.

As a direct consequence of the high biomass concentration obtained in membrane reactors with complete sludge retention, bacteria are facing conditions where the supply of substrate per unit biomass is low [4,5]. At limiting amounts of energy available, it is reasonable to expect that the micro-organisms will preferentially use the carbon sources to satisfy their maintenance energy demands [5,6]. So, if biomass growth only occurs when energy is supplied in excess, in a system where the energy conserved is consumed solely for cell maintenance a maximum biomass concentration will be obtained [7].

The aim of this study was to investigate the microbial community structure and the physiological state of the highly concentrated biomass in a membrane bioreactor aerobically treating municipal wastewater.

2. Materials and methods

2.1. Plant configuration and operating conditions

For details on the plant design and operational modes see [8].

2.2. Sampling

Grab samples for microscopic, phylogenetic and physiological examinations were directly taken from the nitrification tank. For comparison with activated sludge from the conventional wastewater treatment system, grab samples of mixed liquor were collected from the aeration stage of the municipal wastewater treatment plant Berlin-Marienfelde (Berlin, Germany). For fluorescence in situ hybridization, activated sludge samples were fixed according to Manz et al. [9]. Cells were subsequently washed twice with phosphate-buffered saline (PBS; 130 mM NaCl, 7 mM $NaHPO_4$, 3 mM NaH_2PO_4, pH 7.4) and stored in a 1 : 1 mixture of PBS and 96% ethanol at $-20^{\circ}C$. Dispersion and declustering of aggregated flocs in the samples was achieved by sonification using a Branson ultrasonic desintegrator (Sonifier W-250, Heinemann, Schwäbisch-Gmünd, Germany).

2.3. Microscopical examination of sludge structure

Phase contrast and epifluorescence microscopy was performed on a Zeiss Axioplan 2 (Oberkochen, Germany) fitted with appropriate filter sets. In order to determine details of cell morphology, 20 μl of sludge were placed on agar coated glass slides, mounted with a cover slip and examined after cell immobilization. Black and white photomicrographs were taken on Kodak Technical Pan 2415 film.

2.4. Determination of FISH-detectable cells

FISH of fixed activated sludge was performed according to the protocol described by Manz et al. [9]. The oligonucleotide probes used in this study are listed in Table 1. Probe EUB338 was used to quantify members of the domain *Bacteria* with a sufficiently high cellular ribosome content. Taking into account that some bacterial phyla (e.g. members of the phyla Planctomycetales and Verrucomicrobia) are not covered by probe EUB338, one membrane sludge sample (day 351) was additionally analyzed by simultaneous hybridization with equimolar concentrations of probes EUB338, EUB338-II and EUB338-III [10]. The custom synthesized oligodeoxynucleotides were purchased 5′-labelled with Cy3 (Interactiva Biotechnologie, Ulm, Germany). Probe CNIT3 was used as an unlabelled competitor together with probe NIT3 [11]. Staining for

Table 1
Oligonucleotides used and results obtained in this study

Probe	Specificity	FA (%)[a]	Mean % of DAPI-stained cells[b]
EUB338	Domain bacteria	35	38 (±5.4)–49 (±8.4)
EUB338-II	Members of the phylum Planctomycetales	35	38 (±5.1)[c]
EUB338-III	Members of the phylum Verrucomicrobia	35	38 (±5.1)[c]
Non-EUB338	Negative control for non-specific binding	35	n.d.[d]
ARCH915	Domain Archaea, control for cells not detectable with EUB338	30	1 (±0.3)
ALF1b	α-subclass of Proteobacteria and some other bacteria	35	2 (±1.0)–6 (±2.8)
BET42a	β-subclass of Proteobacteria	35	14 (±4.5)–18 (±3.3)
GAM42a	γ-subclass of Proteobacteria	35	1 (±0.5)–2 (±0.9)
SRB385w	Members of the δ-subclass of Proteobacteria, including sulphate-reducing bacteria and some Gram-positive bacteria	35	3 (±1.2)
CF319a	Members of the Cytophaga–Flavobacterium cluster	35	2 (±1.6)
HGC	Gram-positive bacteria with a high G+C DNA content	25	<1
NEU23a	Halophilic and halotolerant members of lithoautotrophic ammonia-oxidizing bacteria	40	n.d.
NSO190	Ammonia-oxidizing β-subclass Proteobacteria	55	n.d.
NSO1225	Ammonia-oxidizing β-subclass Proteobacteria	35	n.d.
NIT2	*Nitrobacter* spp.	40	n.d.
NIT3	*Nitrobacter* spp.	40	n.d.
CNIT3	Unlabelled competitor to ensure the specificity of probe NIT3 when applied simultaneously	40	
NSR1156	Freshwater *Nitrospira* spp.	30	<1
ACA	*Acinetobacter* spp.	35	<1
LDI	*Leptothrix discophora*	35	<1
SNA	*Sphaerotilus natans*	45	<1

[a] Percentage formamide in the hybridization buffer.
[b] Standard deviation given in parenthesis.
[c] Determined by simultaneous application of probes EUB338, EUB338-II and EUB338-III.
[d] Not detectable.

total cell count determination was performed with a $0.5\,\mu g\,ml^{-1}$ DAPI-solution (4,6-diamidino-2-phenylindole; Sigma, Deisenhofen, Germany) for 5 min after hybridization and washing. For each sample and oligonucleotide probe used, at least 10 different, randomly chosen microscopic fields and a minimum of 1000 DAPI-stained cells were enumerated. The relative abundance of specifically hybridized cells was standardized to the total cell count determined by DAPI staining [9]. Statistical evaluations were performed using the Microsoft Excel software program, version 7.0 (Microsoft Corporation).

2.5. Gram staining

For the determination of the Gram staining behavior of the activated sludge cells, the Merck Gram staining kit (Merck, Darmstadt, Germany) was used as recommended by the manufacturer.

2.6. Respiration studies

The oxygen uptake rates of activated sludge samples (300 ml) were measured by insertion of a membrane oxygen electrode equipped with an integrated temperature sensor (TriOximatic EO 200, WTW, Weilheim, Germany), and recording dissolved oxygen concentrations as a function of time. Mixing was provided by a magnetic stirrer and a rubber stopper was used to prevent oxygen transfer. The oxygen consumption rates were determined immediately after the sludge had been taken from the aerated tanks and supplemented with glucose (100 mM) or synthetic wastewater (3.3%; vol/vol), containing 16 g peptone from casein, 11 g beef extract, 3 g urea and 2 g $MgSO_4 \cdot 7H_2O$ per liter distilled water. Control measurements were performed without the addition of external substrates. The oxygen consumption curves were fitted by linear regression with SigmaPlot 2.0 (Jandel Scientific, Erkrath, Germany).

3. Results

3.1. Microscopical examinations

Sludge with a dry matter content of more than $15\,g\,l^{-1}$ showed a dense structure. The morphological composition and the ratio between suspended free and aggregated cells varied over a broad range (Fig. 1). Few common structures of the sludge were found under all conditions in considerable to high numbers. Spherical clusters with a diameter of 10–30 μm were typically observed inside or attached to flocs (Fig. 1a). These colonies consisted of coccoid or short rod-shaped bacteria. The filamentous structures formed by large cocci (Fig. 1b) were also found in the wastewater entering the reactor and were therefore assumed to rather persist in the conventional activated sludge community than to represent indigenous members of the membrane sludge community.

The overall structure of the activated sludge underwent dramatic changes, even under conditions when the input of wastewater and the operating conditions were kept constant over long periods. For the most part of the study the biomass consisted of free suspended single cells along with the dense round colonies described above as the only aggregates (Fig. 1a). Whereas flocs similar to those observed in conventional activated sludge were rarely present, the number of filamentous bacteria could vary from high numbers of 3–4 types of filaments (Fig. 1c) to appearances with the predominance of thin, long filaments that produced high

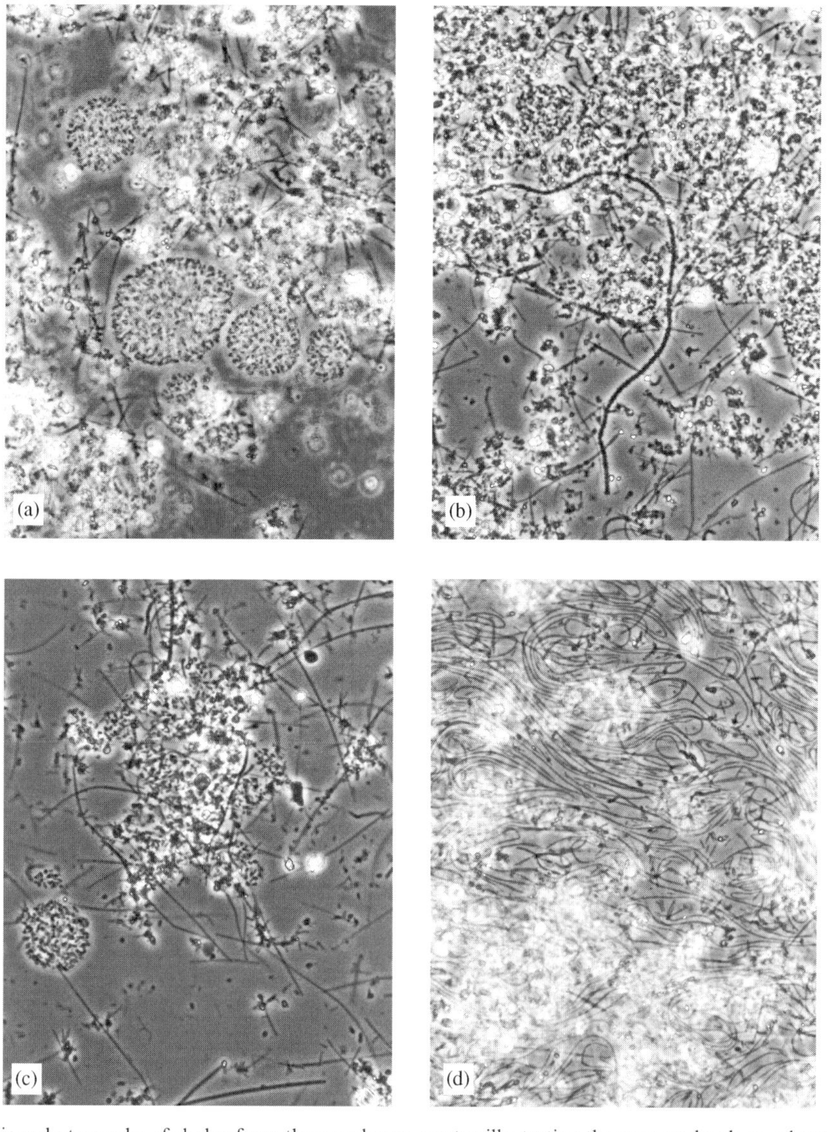

Fig. 1. (a and b) Microphotographs of sludge from the membrane reactor illustrating the commonly observed round, dense colonies and filaments formed of coccoid cells. (c and d) Microphotographs of the membrane reactor demonstrating the variability of the floc structure. Short filaments dominating the flocs and free water and characteristic long, thin and flexible filaments, which were embedded in slime matrices.

amounts of EPS and formed aggregates of a typical pattern (Fig. 1d). At several instances very high numbers of spirochetes were observed, while they were present only in very low numbers most of the time. The number of microscopically detectable protozoa accounted during almost the complete study for less than one protozoon per microscopic field at a 400-fold magnification. In fact, protozoa and metazoa could exclusively be observed during short periods immediately after the operating conditions had been changed.

It is especially noteworthy that the shifts in population composition and biomass structure were never accompanied by changes in the degradation rate, which always remained at about 95% COD removal.

3.2. Detectability of cells by fluorescence in situ hybridization

The detectability of bacteria by FISH is dependent on their ribosomal content and consequently on their physiological state [12]. In the current study, the probe EUB338 was used to assess the overall physiological state of bacterial sludge communities. In sludge samples taken at different days from the membrane filtration reactor, the percentages of cells giving a clear fluorescence signal after hybridization varied between 38% (±5.4%) and 49% (±8.4%) of all DAPI-stained cells (Fig. 2). Additionally performed hybridization experiments with the probes EUB338, EUB338-II and EUB338-III [10] resulted in probe-specific cell counts which were in the same range as observed with probe EUB338 (38%±5.1% compared to 44%±4.4%). In situ probing of activated sludge samples from the conventional wastewater treatment plant Berlin-Marienfelde, receiving the same wastewater as the membrane reactor, yielded 76% (±6.7%) to 78% (±4.9%) of detectable cells (Fig. 2). In order to assess the influence of substrate availability on the FISH-detectable cell counts, sludge loads and MLSS concentrations in the two treatment systems were combined with the results from the hybridization experiments. The MLSS concentration and sludge load in the conventional plant were 2.81 $g\,l^{-1}$ and 0.40 kg COD (kg MLSS)$^{-1}$ d^{-1}, respectively, whereas the MLSS concentration and sludge load in the membrane filtration sludge ranged between 6 and 17.5 $g\,l^{-1}$, and 0.08 and 0.30 kg COD (kg MLSS)$^{-1}$ d^{-1}, respectively. As shown in Fig. 3, the percentages of FISH-detectable cells in the membrane filtration sludge were almost identical and were not affected by changing operating conditions.

Gram staining of membrane filtration sludge samples revealed that less than 10% of bacterial cells were Gram positive. Hybridization with probe HGC, specific for Gram positive bacteria with a high G+C DNA content, resulted in the visualization of less than 1% of cells stained with DAPI. Application of probe ARCH915 indicated that *Archaea* were detectable only at low numbers (1%±0.3%).

3.3. Phylogenetic analysis of the membrane filtration sludge population

FISH analysis was also performed to characterize the overall phylogenetic composition and diversity of the

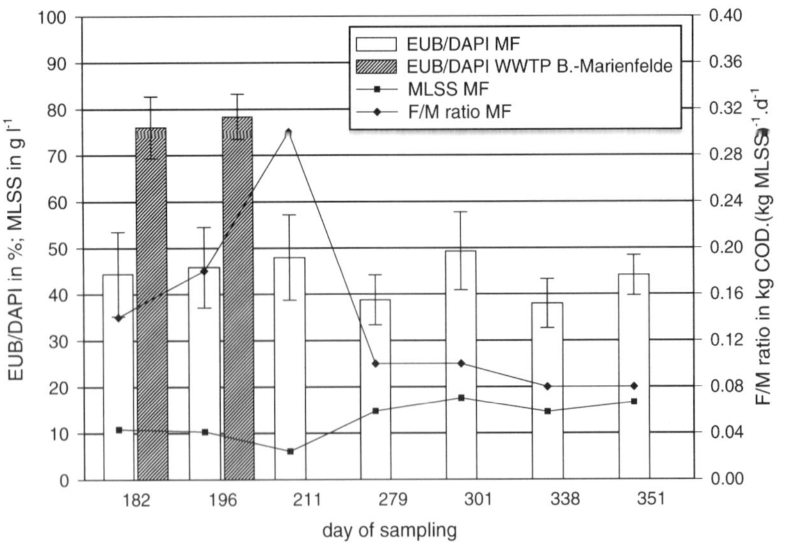

Fig. 2. Percentages of EUB-detectable cells relative to DAPI counts in activated sludge samples taken from the membrane filtration reactor (MF) and the wastewater treatment plant (WWTP) Berlin-Marienfelde. The error bars represent the standard deviation.

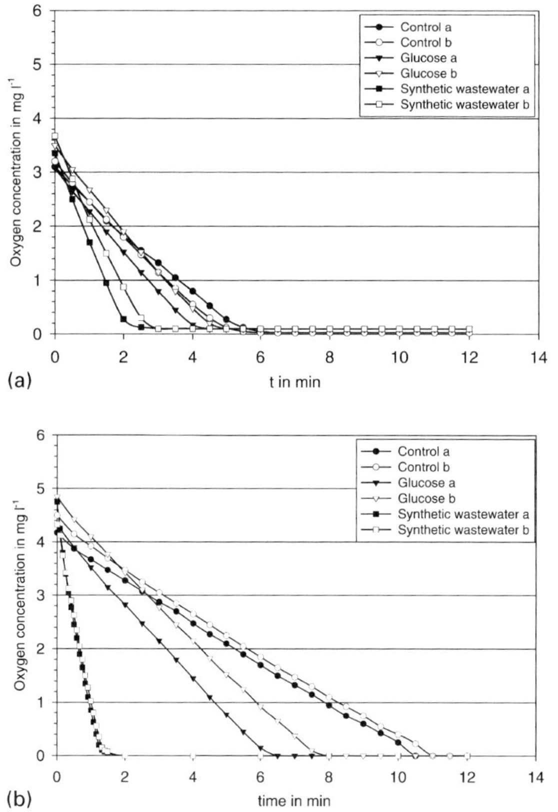

Fig. 3. Exemplary oxygen consumption profiles recorded in activated sludge samples obtained from the conventional WWTP Berlin-Marienfelde (a) and the membrane filtration reactor (b).

microbial population in the membrane bioreactor. The probes applied covered different phylogenetic levels, ranging from group-specific probes (α-, β-, and γ-subclass of Proteobacteria) to probes specific for genera and species frequently shown in activated sludge from conventional wastewater treatment plants (Table 1).

The membrane filtration sludge community detectable by FISH was dominated by members of the β-subclass of Proteobacteria. Cells hybridizing with probe BET42a were characteristic, rod-shaped, aggregated bacteria and accounted for 14% ($\pm$4.5%) to 18% ($\pm$3.3%) of the cells detectable with DAPI (Table 1). Bacteria affiliated to the α- and γ-subclasses of Proteobacteria represented 2% ($\pm$1.0%) to 6% ($\pm$2.8%) and 1% ($\pm$0.5%) to 2% ($\pm$0.9%) of all cells visualized by DAPI staining, respectively. With probe SRB385w, specific for members of the δ-subclass and some members of the α-subclass of Proteobacteria, 3% ($\pm$1.1%) of the DAPI-stained cells could be detected. Members of the Cytophaga–Flavobacteria cluster, detected with probe CF319a, accounted for 2% ($\pm$1.6%) of the total number of cells stained with DAPI.

Species-specific probes for bacteria repeatedly found in activated sludge (*Sphaerotilus natans, Leptothrix discophora* and *Acinetobacter* spp.), hybridized to single, rod-shaped cells, but never accounted for more than 1% of the DAPI-stained cells. By screening for ammonia- and nitrite-oxidizing bacteria, no cells hybridizing with probes NSO190 and NSO1225, both specific for *Nitrosomonas* sp. and *Nitrosospira* sp., probe NEU23a, specific for most halophilic and halotolerant ammonia-oxidizing bacteria, and probes NIT2 and NIT3, specific for *Nitrobacter* spp. were detected. In situ hybridization with the probe NSR1156 revealed that less than 1% of the total cell counts could be assigned to the genus *Nitrospira*.

3.4. Oxygen consumption rates

Oxygen consumption rates, which were considered to be equivalent to the overall metabolic activity of the sludge community, were determined to elucidate the physiological state and metabolic activity of the bacteria present in the membrane filtration sludge. The oxygen uptake rates were recorded immediately after sampling to prevent the sludge community from physiological changes. There was no further aeration before starting the measurements, but the initial oxygen concentrations slightly increased due to oxygen transfer while the sludge was transferred into Erlenmeyer flasks. Activated sludge from the conventional wastewater treatment plant Berlin-Marienfelde was investigated in the same way as the sludge from the membrane reactor. Oxygen uptake profiles for the conventional activated sludge and membrane filtration sludge samples are shown in Figs. 3a and b. Specific oxygen consumption rates as determined by linear regression are presented in Table 2. Without addition of supplementary energy sources, sludge originating from the conventional wastewater treatment plant respired the remaining substrates at specific oxygen consumption rates of 0.280 mg O_2 (g MLVSS)$^{-1}$ min^{-1}. After addition of different easily degradable carbon sources, the specific rates of oxygen consumption in the conventional activated sludge did not change notably (Fig. 3a). The specific respiration rates increased to 0.355 (+27%) and 0.668 (+139%) mg O_2 (g MLVSS)$^{-1}$ min^{-1} when glucose or synthetic wastewater was used as an additional carbon source, respectively (Table 2). If no further carbon sources were added, sludge samples taken from the membrane reactor consumed the dissolved oxygen at specific rates of 0.032–0.036 mg O_2 (g MLVSS)$^{-1}$ min^{-1} (Fig. 3b and Table 2). Despite the comparatively high MLVSS concentrations, the specific oxygen uptake rates for the oxidation of the residual substrates in the membrane filtration sludge were about eight times lower than those found in unamended conventional activated sludge controls. After the addition of the same carbon sources,

Table 2
Specific oxygen consumption rates in activated sludge samples from the membrane filtration reactor (MF) and the wastewater treatment plant (WWTP) Berlin-Marienfelde without and at presence of easily degradable carbon sources

	Sample (day of operation)			
	WWTP Berlin-Marienfelde	MF (308)	MF (338)	MF (351)
Control	0.280[a,b]	0.034	0.036	0.032
Glucose	0.355	0.064	0.062	0.056
Synthetic wastewater	0.668	0.391	0.324	0.303
Sludge temperature (°C)	14	24	23	22

[a] Expressed in mg O_2 (g MLVSS)$^{-1}$ min^{-1}.
[b] Mean values obtained from two experiments.

the oxygen consumption immediately increased resulting in specific respiration rates of 0.056–0.064 mg O_2 (g MLVSS)$^{-1}$ min^{-1} in the case of glucose, and 0.303–0.391 mg O_2 (g MLVSS)$^{-1}$ min^{-1} after addition of synthetic wastewater (Table 2). Thus, compared to values obtained in the unamended controls, a more pronounced increase in respiration activity could be observed in the membrane filtration sludge after the addition of glucose (+80%) and synthetic wastewater (+900%). The sludge temperatures were determined to 23°C (±1°C) in the membrane filtration sludge and 14°C in the conventional activated sludge.

4. Discussion

The application of integrated microfiltration membranes in activated sludge systems has a variety of consequences with respect to the process features, including the virtually bacteria free effluent, no excess secondary sludge (with exception of non-hydrolyzable particles) and stable performance of the reactor regarding carbon and nitrogen removal. In contrary to the models applied in conventional activated sludge systems, the complete retention of biomass implicates modifications for the biology in membrane reactors. Firstly, the biomass retention results in higher biomass concentrations compared to conventional activated sludge processes, where the sludge content mainly depends on the performance of the secondary clarifier. Secondly, the washing out of slow growing bacteria from the bioreactor can be avoided.

One of the major limitations of conventional activated sludge systems is bulking provoked by high numbers of filamentous bacteria which cause poor settling sludge flocs. As a consequence, biomass is permanently lost from the secondary clarifier and floc-associated pathogens reduce the hygienical quality of the effluent. In the membrane bioreactor examined, high numbers of filamentous bacteria were present at several periods without any adverse effects for the process. The application of a 0.2 μm pore sized membrane resulted in an effluent water quality, which meets the hygienical requirements for bathing water, including a complete retention of cysts of parasites (*Cryptosporidium*, *Giardia*, *Microsporidia*) and eggs of parasitic worms (data not shown).

Compared to conventional activated sludge systems the micro-organisms in the membrane filtration reactor are exposed to completely different environmental conditions. Most importantly, grazing by protozoa and metazoa did not occur and sludge was not removed except for sampling. Under these conditions the key question arises how a stable biomass concentration was maintained, as soon as zero net biomass growth and sludge production was observed. In a system without predation and biomass removal, the net biomass production is limited by (i) the amount of energy provided, (ii) the maintenance energy demand and (iii) cell decay processes [6,13]. From the maintenance energy concept introduced by Pirt [14] it can be concluded that the growth rates of the accumulating bacteria decrease along with the substrate availability [7]. Apparent zero net biomass generation can be based on two different physiological concepts: according to the growth–death concept, the substrate available per unit biomass allows one part of the cells to grow, the other part is exposed to decay due to the loss of cell viability. In this context, growth rates and death rates of micro-organisms may be correlated as shown by Kurland and Mikkola [15]. Slow cell growth would then result in correspondingly low death rates. The growth–death concept also includes growth on secondary substrates originating from cell lysis. The maintenance energy concept assumes that micro-organisms primarily utilize the energy supplied to satisfy their maintenance energy requirements. Maintenance functions include the general turnover of cell materials, continuous replacement of proteins and RNA, osmotic work to maintain concentration gradients and cell motility [5,6]. Only if energy is supplied in excess, bacteria are able to grow. When the maximum biomass concentration is reached, the maintenance energy requirements of the biomass will just be met by the limiting nutrient supply [7] and the terms for growth and die-off approach zero.

The present study intended to verify the model concept most useful to be applied to describe the physiological state of the sludge consortium in the membrane filtration reactor. For this purpose, two experimental strategies using different parameters for the assessment of the physiological state of bacterial cells were applied to compare the characteristics of conventional sludge and membrane filtration sludge.

From comparison of the oxygen uptake rates upon addition of substrates it is concluded that the cells in the membrane reactor are not limited by oxygen but that a limitation for organic carbon is more likely. The easily degradable substrates added were used by both sludge populations without lag-phase. Despite the amount of energy supplied was higher with glucose (COD: $19{,}200\,mg\,l^{-1}$), the increase in oxygen uptake rates was more pronounced when a complex mixture of substrates (synthetic wastewater, COD: $400\,mg\,l^{-1}$) was added. Although the temperature in the conventional treatment plant was 9°C lower than in the membrane filtration sludge, the oxygen consumption rates of the conventional sludge were always significantly higher.

FISH analysis revealed β-subclass Proteobacteria to constitute the predominant group of bacteria within the sludge community. A numerical domination of β-Proteobacteria has been observed in many natural and man-made habitats, such as drinking water [16] and activated sludge [17,9]. Screening with probes specific for chemolithoautotrophic ammonia-oxidizing bacteria showed neither *Nitrosomonas* nor *Nitrosospira*, which are commonly found in conventional systems with nitrification activity [11], to be present in the membrane filtration sludge. The measured nitrification activity could therefore be caused by presently unknown lithoautotrophic ammonia-oxidizing species or heterotrophic nitrifiers. As previously found in conventional sewage systems [18], no *Nitrobacter* spp. could be detected in the membrane filtration system. Recently, the importance of the genus *Nitrospira* for nitrifying systems has been demonstrated [19]. In addition, Wagner et al. [18] showed that nitrite-oxidation in activated sludge has not to be related to *Nitrobacter* spp. The in situ detection of *Nitrospira* sp. in this study confirms that *Nitrospira* spp. might be of more general importance for nitrite oxidation than the genus *Nitrobacter* [19].

Besides providing phylogenetic information, FISH can be used to assess the physiological state of single bacterial cells [16]. In conventional activated sludge, substrate degradation is coupled to biomass growth, i.e. the cells gain excess energy for anabolic reactions. As a consequence of the high cellular rRNA contents, generally 70–90% of the cells stained by DAPI were also detectable by FISH [17,9]. The results obtained for the treatment plant Berlin-Marienfelde were in agreement with those previously found in conventional activated sludges. Similarly low amounts of hybridized cells as found in the membrane filtration sludge have been reported only once for a low-load stage of two successive activated sludge basins [17]. In the membrane reactor sludge less than 50% of the cells contained enough rRNA molecules for visualization by FISH. Cells with low rRNA contents were often associated with dormant, starved, or very slowly growing cells, all representing stages of low metabolic activity. However, the substrate elimination in the membrane filtration reactor occurred at superior levels and there was no indication for an accumulation of substances being recalcitrant towards degradation, either from the inflow [8] or from decayed cells. This indicates that bacteria with low ribosome contents also actively participate in metabolic transformations. Recently, it was shown that several members of the two environmentally important bacterial phyla Planctomycetales and Verrucomicrobia could not be detected by probe EUB338 [10]. These bacteria may account for an unknown fraction of DAPI-stained cells which are not visualized by in situ hybridization. In our study, however, additional hybridization experiments with the probe set recommended by Daims et al. [10] did not result in increasing probe-specific cell counts, indicating that neither Planctomycetales nor Verrucomicrobia were present in significant numbers. Gram staining and application of the *Archaea*-specific probe ARCH519 furthermore suggest that the low numbers of FISH-detectable bacteria were caused neither by the presence of Gram positive bacteria nor by *Archaea*.

Since it was shown that carbon-starved, non-growing bacteria could retain large numbers of ribosomes for several days [20], it is not possible to conclude from high signal strength on a high growth rate. Nevertheless, the presence of a minimum number of ribosomes seems to be a prerequisite for cell division. It is therefore hypothesized that the majority of the cells in the membrane reactor are not in a physiological state characteristic for growth. It should be kept in mind, however, that cell decay must occur at a certain rate in the membrane reactor, since new biomass was permanently introduced into the reactor with the influent, and grazing organisms were not present most of the time.

Most kinetic models presently used for the description of the dynamics of microbial populations in wastewater treatment have been derived from the conventional activated sludge process. In this process, the rate of biomass growth is controlled by the wash out of cells and discharge of secondary sludge, as well as the grazing activity of protozoans, rotifers and nematodes. The latter mechanism is often considered to be of minor relevance and was usually subsumed under the decay/

lysis of biomass in the activated sludge models [13]. There is, however, growing evidence that grazing effects are much more relevant for the control of the biomass density than biomass reduction related to dying bacteria [13]. In the investigated membrane reactor, where both mechanisms, wash-out and grazing, are not relevant, the maintenance energy requirement of the microbial population is obviously more important for the finally attained biomass content than in conventional activated sludge systems. Further research is needed before a unifying model for membrane reactors can be formulated.

5. Conclusions

- In a membrane reactor treating municipal wastewater at high biomass concentrations the bacterial population was able to mineralize the substrates at a high and stable rate.
- The membrane sludge population was shown to be more substrate-limited than conventional activated sludge treating the same wastewater.
- Microbial populations from the membrane reactor were found to contain significantly less bacterial cells detectable by FISH compared to conventional activated sludge.
- The system was virtually free of grazing organisms, and therefore lacking one of the most important mechanisms responsible for cell elimination and ecological selection in conventional activated sludge.
- The majority of cells were assumed to reside in a physiological state not allowing for cell division. However, these non-growing cells are thought to participate in the degradation processes to satisfy their maintenance energy requirements.

References

[1] Ternes TA. Occurrence of drugs in German sewage plants and rivers. Water Res 1998;32:3245–60.

[2] Prüss A. Review of epidemiological studies on health effects from exposure to recreational water. Int J Epidemiol 1998;27:1–9.

[3] Krauth K-H, Staab KF. Pressurized bioreactor with membrane filtration for wastewater treatment. Water Res 1993;27:405–11.

[4] Yamamoto K, Hiasa M, Mahmood T, Matsuo T. Direct solid-liquid separation using hollow fiber membrane in an activated sludge aeration tank. Water Sci Tech 1989;21: 43–54.

[5] Muller EB, Stouthamer AH, van Verseveld HW, Eikelboom DH. Aerobic domestic wastewater treatment in a pilot plant with complete sludge retention by cross-flow filtration. Water Res 1995;29:1179–89.

[6] Low EW, Chase HA. The effect of maintenance energy requirements on biomass production during wastewater treatment. Water Res 1999;33:847–53.

[7] Beyeler W, Rogers PL, Fiechter A. A simple technique for the direct determination of maintenance energy coefficient: an example with *Zymomonas mobilis*. Appl Microbiol Biotechnol 1984;19:277–80.

[8] Rosenberger S, Krüger U, Witzig R, Manz W, Szewzyk U, Kraume M. Performance of a bioreactor with submerged membranes for aerobic treatment of municipal waste water. Water Res 2002;36(2):413–20.

[9] Manz W, Wagner M, Amann R, Schleifer K-H. In situ characterization of the microbial consortia active in two wastewater treatment plants. Water Res 1994;28:1715–23.

[10] Daims H, Brühl A, Amann R, Schleifer K-H, Wagner M. The domain-specific probe EUB338 is insufficient for the detection of all bacteria: development and evaluation of a more comprehensive probe set. Syst Appl Microbiol 1999;22:434–44.

[11] Wagner M, Rath G, Amann R, Koops H-P, Schleifer K-H. In situ identification of ammonia-oxidizing bacteria. Syst Appl Microbiol 1995;18:251–64.

[12] Amann RI, Ludwig W, Schleifer K-H. Phylogenetic identification and in situ detection of individual microbial cells without cultivation. Microb Rev 1995;59:143–69.

[13] Van Loosdrecht MCM, Henze M. Maintenance, endogeneous respiration, lysis, decay and predation. Water Sci Tech 1999;39(1):107–17.

[14] Pirt SJ. The maintenance energy of bacteria in growing cultures. Proc R Soc Ser B 1965;163:224–31.

[15] Kurland CG, Mikkola R. The impact of nutritional state on the microevaluation of ribosomes. In: Kjelleberg S, editor. Starvation in bacteria. New York: Plenum Press, 1993. p. 225–37.

[16] Kalmbach S, Manz W, Szewzyk U. Dynamics of biofilm formation in drinking water: phylogenetic affiliation and metabolic potential of single cells assessed by formazan reduction and in situ. FEMS Microbiol Ecol 1997;22: 265–79.

[17] Wagner M, Amann R, Lemmer H, Schleifer K-H. Probing activated sludge with oligonucleotides specific for Proteobacteria: inadequacy of culture-dependent methods for describing microbial community structures. Appl Environ Microbiol 1993;59:1520–5.

[18] Wagner M, Rath G, Koops H-P, Flood J, Amann R. In situ analysis of nitrifying bacteria in sewage treatment plants. Water Sci Tech 1996;34(1-2):237–44.

[19] Schramm A, De Beer D, Wagner M, Amann R. Identification and activities in situ of *Nitrosospira* and *Nitrospira* spp. as dominant poulations in a nitrifying fluidized bed reactor. Appl Environ Microbiol 1998;64:3480–5.

[20] Flärdh K, Cohen PS, Kjelleberg S. Ribosomes exist in large excess over the apparent demand for protein synthesis during carbon starvation in marine *Vibrio* sp. strain CCUG 15956. 1992.

PERGAMON

Water Research 36 (2002) 403–412

WATER RESEARCH

www.elsevier.com/locate/watres

Characterization of denitrifying phosphate-accumulating organisms cultivated under different electron acceptor conditions using polymerase chain reaction-denaturing gradient gel electrophoresis assay

Johwan Ahn, Tomotaka Daidou, Satoshi Tsuneda*, Akira Hirata

Department of Chemical Engineering, Waseda University, 3-4-1 Ohkubo, Shinjuku-ku, Tokyo 169-8555, Japan

Received 15 June 2000; received in revised form 4 January 2001; accepted 17 April 2001

Abstract

To investigate the characteristics and the microbial diversity of denitrifying phosphate-accumulating organisms (DNPAOs) that are capable of conducting enhanced biological phosphorus removal (EBPR) using nitrate as electron acceptor, three sequencing batch reactors were operated under three different electron acceptor conditions, i.e., only oxygen, oxygen together with nitrate and only nitrate. Based on the chemical analysis concerning the biochemical transformation of each reactor, it was found that phosphate-accumulating organisms responsible for EBPR consisted of at least three populations including DNPAOs, and that the microbial community structure was changed according to the electron acceptor conditions. Also, the sludge cultivated with oxygen together with nitrate showed a drastic increase in the amount of phosphorus uptake under anoxic conditions, which suggested that a proportion of DNPAOs capable of utilizing nitrate under aerobic conditions were present.

On the other hand, the change in microbial community structure depending on the type of electron acceptor was demonstrated by the analysis of the results of denaturing gradient gel electrophoresis of polymerase chain reaction-amplified 16S ribosomal DNA fragments. It was found that the bacteria commonly contained in all the reactors were *Rhodocyclus* sp. (96% identity) and *Dechlorimonas* sp. (97% identity) that belonged to the beta subclass of Proteobacteria on the basis of the analysis of the sequence excised from DGGE bands and the determination of phylogenetic affiliation. However, only the presence of *Rhodocyclus* sp. in all the reactors was demonstrated by fluorescent in situ hybridization analysis. © 2002 Elsevier Science Ltd. All rights reserved.

Keywords: 16S ribosomal DNA; Denaturing gradient gel electrophoresis; Denitrifying phosphate-accumulating organisms; Enhanced biological phosphorus removal; Fluorescent in situ hybridization; Phosphate-accumulating organisms; Polymerase chain reaction

1. Introduction

Biological nutrient removal (BNR) processes have been widely used to treat wastewater containing nutrient materials, i.e., nitrogen and phosphorus, as well as chemical oxygen demand (COD) over the past several decades because of their economic advantage compared with chemical treatment methods. Since phosphorus is the limiting nutrient in algal blooms, the enhanced biological phosphorus removal (EBPR) process characterized by phosphate-accumulating organisms (PAOs) has been regarded as a means for preventing eutrophication. However, EBPR processes combined with biological nitrogen removal process have a drawback in that

*Corresponding author. Tel.: +81-352863210; fax: +81-332093680.

E-mail address: stsuneda@mn.waseda.ac.jp (S. Tsuneda).

PII: S0043-1354(01)00222-6

COD acts as a limiting factor for phosphorus release and denitrification at a low C/N ratio.

If that is the case, the most appropriate solution to the problem associated with the limitation effect of COD is the introduction of denitrifying phosphate-accumulating organisms (DNPAOs) into the BNR process, which are capable of utilizing nitrate as electron acceptor to simultaneously remove phosphorus and nitrogen from wastewater [1–6]. Another advantage of employing DNPAOs in the BNR process is that the use of nitrate rather than oxygen as electron acceptor leads to a smaller amount of sludge produced as well as the efficient use of COD. Thus, nitrate, which is produced during the course of nitrification under aerobic conditions, has been considered as another electron acceptor for simultaneously removing nitrogen and phosphorus without additional supplementation of an extracellular carbon substrate [7].

Recently, in order to consider the proportion of PAOs capable of anoxic phosphorus uptake by using nitrate instead of oxygen as electron acceptor, the denitrifying activity of DNPAOs were introduced into the Activated Sludge Model (ASM) No. 2d by the IAWQ task group [8]. Nevertheless, a detailed understanding of DNPAOs from a microbiological point of view is lacking. This is mainly because DNPAOs that participate in the key chemical transformations for EBPR are so far nonculturable using classical culture techniques. A few studies have reported the detection of PAOs that belong to the beta subclass of Proteobacteria by means of several molecular techniques [9–13]. However, these studies have been limited to PAOs with high phosphorus-accumulating ability under anaerobic/aerobic conditions.

Thus, to determine DNPAOs that are responsible for anoxic phosphorus removal in the EBPR process, we adapted the polymerase chain reaction (PCR)-denaturing gradient gel electrophoresis (DGGE) method that has been widely recognized as a powerful tool in the study of microbial ecology.

Denaturing gradient gel electrophoresis is based on the electrophoresis of PCR-amplified 16S ribosomal DNA (rDNA) fragments in polyacrylamide gels containing a linearly increasing denaturant gradient [14]. Different DGGE bands are separated depending on the melting behavior of the PCR products with the same length but with different sequences, which correspond to the species. Therefore, PCR–DGGE assay allows us to analyze the microbial composition and diversity of a given system without the need to isolate individual species [15,16]. It is particularly effective in analyzing the microbial community structure of nonculturable environmental samples such as microorganisms responsible for EBPR. Furthermore, sequence analysis of the bands excised from DGGE gel provides a basis for the phylogenetic identification of the bacterium corresponding to each band and the design of 16S ribosomal RNA (rRNA)-targeted oligonucleotide probes for fluorescent in situ hybridization (FISH) analysis, even though these bacteria are nonculturable.

The purpose of this study is to characterize DNPAOs cultivated under different electron acceptor conditions, and on the basis of the results, to analyze the change in microbial community structure, phylogenetic affiliation and identification of DNPAOs that perform EBPR by using PCR-DGGE and FISH.

2. Materials and methods

2.1. Reactor operation

The operation of three sequencing batch reactors (SBRs) each with 1.0 l working volume, which were supplied with three different electron acceptors, i.e., oxygen together with nitrate (RAN reactor), without nitrate (RAA reactor) and only nitrate (RNN reactor), was separately conducted for 30 days to obtain PAOs responsible for EBPR. As shown at Table 1, RAN and RAA reactors were operated under anaerobic/aerobic conditions, while the RNN reactor was operated under anaerobic/anoxic conditions.

Activated sludge withdrawn from an aerobic basin of a municipal wastewater treatment plant located in Tokyo was inoculated before operating the reactor. Prior to inoculation, the anoxic phosphorus uptake ability of this sludge was evaluated using nitrate as electron acceptor. As a result, total organic carbon (TOC) was completely taken up and phosphorus was released under anaerobic conditions, and phosphorus uptake using nitrate as electron acceptor was observed under subsequent anoxic conditions.

All the reactors were operated with a cycle of 8 h, consisting of a 20-min filling phase, a 90-min anaerobic phase, a 315-min aerobic or anoxic phase, a 30 min

Table 1
Conditions for reactor operation

Reactor	RAN	RAA	RNN
Electron acceptor	Air + nitrate[a]	Air[b]	Nitrate[c]

[a] Nitrate was supplied by nitrification of ammonium nitrogen under aerobic conditions.

[b] Nitrification was controlled by *n*-allylthiourea, which is known as a nitrification-inhibiting agent, to inhibit exposure of the sludge to nitrate or nitrite produced under aerobic conditions.

[c] Nitrate solution (NO_3–N: 50 mg) was supplied at the beginning of anoxic conditions during the first 100 min.

settling phase and a 25-min withdrawing phase. After one cycle, 0.5 l of the effluent was exchanged with an influent of the same amount, and thus 16 h of hydraulic retention time (HRT) was maintained. At the end of one cycle, 17 ml excess activated sludge was also wasted to maintain a mean cell retention time (MCRT) of 20 days. The pH was maintained between 6.9 and 7.1 during the entire operation period. The relevant components were periodically measured to monitor EBPR performance during this study.

2.2. Synthetic wastewater

The basic medium contained the following compounds per liter: 1.7 g $CH_3COONa \cdot 3H_2O$, 176 mg KH_2PO_4, 471 mg $(NH_4)_2SO_4$, 28 mg $CaCl_2 \cdot 2H_2O$, 180 mg $MgSO_4 \cdot 7H_2O$ and 0.6 ml of nutrient solution. The nutrient solution consisted of the following compounds per liter: 1.5 g $FeCl_3 \cdot 6H_2O$, 0.15 g H_3BO_3, 0.03 g $CuSO_4 \cdot 5H_2O$, 0.18 g KI, 0.12 g $MnCl_2 \cdot 4H_2O$, 0.06 g $Na_2MoO_4 \cdot 2H_2O$, 0.12 g $ZnSO_4 \cdot 7H_2O$, 0.15 g $CoCl_2 \cdot 6H_2O$, and 10 g EDTA.

2.3. Chemical analysis

The analysis of mixed liquor suspended solids (MLSS) and phosphate was performed in accordance with Standard Methods for the Examination of Water and Wastewater [17]. Nitrate was determined using a high-performance liquid chromatograph equipped with an anion column (IC-Anion-PW, Tosoh, Japan) and with an ultraviolet detector (UV-8011, Tosoh, Japan). TOC was measured by a TOC analyzer (TOC-500, Shimadzu, Japan).

2.4. Batch experiments

Activated sludge was taken from the reactor at the end of aerobic or anoxic conditions at 30 days, and was immediately washed twice in the nutrient solution (pH 7.0) that did not contain the basic medium. The activated sludge prepared was divided into three parts. Two parts were incubated with the synthetic wastewater with and without *n*-allylthiourea, respectively, under anaerobic conditions, and then were exposed to aerobic conditions. Another part was exposed to anoxic conditions after being incubated with the synthetic wastewater without *n*-allylthiourea under anaerobic conditions. For anoxic conditions, nitrate solution (NO_3–N: 50 mg) was added at the beginning. In this way, the batch experiments using the activated sludges taken from all the reactors were carried out during 30-min anaerobic conditions and 315-min subsequent aerobic or anoxic conditions, and phosphorus uptake ability under different electron acceptor conditions was investigated and evaluated.

2.5. PCR–DGGE analysis

2.5.1. Sampling and preparation

Sludge samples were collected from each reactor at the end of aerobic or anoxic conditions, immediately concentrated by centrifugation at 8500 rpm (CF15R, Hitachi, Japan) for 5 min at 4°C, and then vortexed with 3 volumes (vol/wt) of TE buffer (10 mM Tris–HCl, 1 mM EDTA, pH 8.0) for 5 min. The preparation of the sludge samples as described above was repeated three times, and then stored at −20°C.

2.5.2. DNA extraction

The sludge samples prepared were suspended in 5 volumes (vol/wt) of SL buffer (10% sucrose, 0.7 M NaCl, 40 mM Na_2EDTA, 50 mM Tris–HCl, pH 8.5) containing lysozyme of 1 mg ml^{-1}, and then subjected to ultrasonication for 30 s. After centrifugation for 10 min at 5500 rpm, the aqueous phase was recovered. As described above, the recovery of the aqueous phase was repeated three times.

The aqueous phase recovered was incubated with 0.01 volume of 1% (wt/vol) proteinase K, with 0.05 volume of 10% (wt/vol) sodium dodecylsufate, followed by 0.1 volume of 10% (wt/vol) HTMAB, each for 30 min at 55°C. After centrifugation for 10 min at 10,000 rpm, DNA from the aqueous phase that was transferred to a new tube was extracted with an equal volume of PCI solution (phenol-chloroform-isoamyl alcohol = 25 : 24 : 1, vol/vol/vol) and CIA solution (chloroform-isoamyl alcohol = 24 : 1, vol/vol), and then precipitated with 0.1 volume of 3 M sodium acetate (pH 5.2) and 2.5 volumes of absolute ethanol. The DNA pellet was obtained by washing with 70% (vol/vol) ethanol, and drying in a desiccator. Finally, the dried pellet was resuspended in TE buffer and stored at 4°C.

2.5.3. PCR amplification

The region corresponding to positions 341 and 907 in the 16S rDNA of Escherichia coli was PCR-amplified using the forward primer EUB341f (5′-CCTACGG-GAGGCAGCAG-3′) with a GC clamp (5′-CGCCC GCCGCGCGGCGGCGGGCGGGGCGGCACGGG-GGG-3′) at the 5′ end to stabilize the melting behavior of the DNA fragments and the reverse primer UNIV907r (5′-CCCCGTCAATTCCTTTGAGTTT-3′). PCR amplification was conducted in an automated thermal cycler (GeneAmp PCR System 9700, Applied Biosystems, USA) using the following protocol: initial denaturation for 2 min at 94°C and 30 cycles of denaturation for 1 min at 94°C, annealing for 1 min at 53°C, extension for 1 min at 72°C, followed by a final extension for 2 min at 72°C. PCR mixtures had a final volume of 40 μl which contained 4 μl of 10 × PCR buffer, 0.6 mM $MgSO_4$, 0.2 mM of each primer, 0.08 mM dNTP, 20 pg template and 1 U polymerase. After PCR amplification, PCR products were electrophoresed on 2% (wt/vol) agarose gel in 1 × TAE for 70 min for 50 V, and then checked with ethidium bromide staining before being loaded onto a denaturing polyacrylamide gel.

2.5.4. Electrophoresis

DGGE was carried out using the Dcode™ Universal Mutation Detection System (BioRad, USA) in accordance with the manufacturer's instructions. PCR products were electrophoresed in 1 × TAE buffer (40 mM Tris–HCl, 40 mM CH_3COOH, 1 mM EDTA, pH 8.0) for 240 min at 130 V and 60°C on polyacrylamide gel containing a linear gradient ranging from 30% to 50% denaturant. After electrophoresis, polyacrylamide gel was stained with SYBR Green I (FMC Bioproducts, No. 50513) for 30 min, and then visualized on UV transilluminator (CSF-35BC, Cosmobio, Japan) with a digital camera (DC120 ZOOM, Kodak, USA).

2.5.5. Sequencing

The two bands were excised from DGGE polyacrylamide gel for 16S rDNA fragment sequencing. DNA fragments from the bands excised were purified using a QIAEX II Gel Extraction Kit (Qiagen, USA), and then PCR-amplified with the forward primer EUB341f without a GC clamp and the reverse primer UNIV907r. After PCR amplification, the PCR products were purified using a QIAquick PCR Purification Kit (Qiagen, USA).

Both strands of the purified PCR products were sequenced with primers EUB341f, 514F (5′-GTGCCAGCAGCCGCGG-3′), 786F (5′-ATTAGATACCCTGGTA-3′), 530R (5′-ACCGCGGCTGCTGGC-3′), 802R (5′-CTACCAGGGTATCTAAT-3′) and UNIV907r by ABI PRISM Big Terminator Cycle Sequencing Kit (Applied Biosystems, USA) and ABI PRISM 377 automated DNA sequencer (Applied Biosystems, USA) in accordance with the manufacturer's instructions.

2.6. FISH analysis

2.6.1. Oligonucleotide probes

16S rRNA-targeted oligonucleotide probes EUB338 (5′-GCTGCCTCCCGTAGGAGT-3′) [18] labeled with CY3 and PAO846 (5′-GTTAGCTACGGCACTAAAAGG-3′) [12] labeled with fluorescein isothiocyanate (FITC) were prepared for in situ detection of *Rhodocyclus* sp. that belonged to beta subclass of Proteobacteria.

On the other hand, since there are no oligonucleotide probes available for in situ detection of *Dechlorimonas* sp., probe DCM433 (5′-GTATTAACCCATGCGATTTC-3′) was originally designed based on sequences accessible in the 16S rRNA database and checked with Ribosomal Database Project (RDP). Based on the check of the number of base pair mismatches in the RDP with probe DMC433, only three organisms with zero (*Dechlorimonas SIUL str.* and *Dechlorimonas MissR str.*) or one (*Ferribacterium limneticum str. cda-1*) bp mismatch were confirmed, because it was very specifically designed for a targeted organism. However, there are no pure cultures whose 16S rRNA can be bound with probe DMC433. Thus, hybridization using the sludge samples instead of pure cultures was conducted under the various ranges of hybridization temperature, wash temperature and formamide concentration for optimizing in situ conditions and for verifying probe specificity with respect to targeted organism.

2.6.2. In situ hybridization

Sludge samples were fixed in 4% paraformaldehyde-phosphate-buffered saline (PBS) solution (PBS; 137 mM NaCl, 8.1 mM Na_2HPO_4, 2.68 mM KCl, 1.47 mM KH_2PO_4, pH 7.2) for 2 h at 4°C. The fixed sludge samples were washed twice in PBS, resuspended in PBS–ethanol solution (1 : 1, vol/vol), and then stored at −20°C. Hybridization was performed according to the standard protocol described by Amann [19]. The fixed cell suspensions prepared above were spotted onto a hybridization slide coated with gelatin and allowed to air-dry. The fixed cell suspension on the hybridization slide was then dehydrated by washing the slide in 50%, 80% and 98% (vol/vol) ethanol for 3 min each. The slide was allowed to air-dry, and then stored in a desiccator prior to hybridization.

Hybridization of the sludge samples was performed for 2.5 h at 46°C in a hybridization buffer (0.9 M NaCl, 20 mM Tris–HCl, 0.01% sodium dodecyl sulfate, 20% (EUB388) and 35% (PAO846) formamide, pH 7.2). Hybridization with a probe requiring a high stringency was performed followed by hybridization with a probe requiring a low stringency. Hybridization was followed by a stringent washing step for 20 min at 48°C in a washing buffer (40 mM NaCl, 0.01% sodium dodecyl sulfate, 20 mM Tris–HCl, pH 7.2). The washing buffer was removed by rinsing the slides with distilled water, and then the slides were immediately dried. Finally, the hybridization slide was mounted using Fluorguard Antifade Reagent Kit (BioRad, USA), and then observed under a confocal scanning laser microscope (TCS NT, Leica, Germany).

3. Results and discussion

3.1. Performance of EBPR

The three SBRs (RAN, RAA and RNN reactors) achieved EBPR under three different electron acceptor conditions. A typical key biochemical transformation responsible for EBPR was observed during the operation period, which strongly suggested that PAOs were predominant in all the reactors used in this study.

As can be seen in Fig. 1, the reactors showed significant differences in the amount of phosphorus uptake per MLSS according to the type of electron acceptor used. The RAN reactor having oxygen together with nitrate produced by nitrification showed the highest amount of phosphorus uptake among all the three SBRs. Moreover, the amount of phosphorus uptake per MLSS in the RAN reactor gradually increased with the operation period and reached 17 mg-P g^{-1}-MLSS after 30 days, which corresponded to almost twice that in the RNN reactor provided with only nitrate. In contrast, the RNN reactor showed a slight decrease in the amount of phosphorus uptake per MLSS at the beginning of the operation period, while the RAA reactor showed no significant change. According to several studies, the amount of energy produced during oxidative phosphorylation with nitrate is approximately 40% lower than that with oxygen [20,21]. Therefore, the present result suggests that the efficiency of phosphorus uptake per MLSS strongly depends on the type of electron acceptor together with the microbial community structure associated with PAOs.

On the other hand, comparison of the amount of phosphorus uptake per MLSS between RAN and RAA reactors reveals that the presence of nitrate together with oxygen promotes the amount of phosphorus uptake per MLSS. This suggests that DNPAOs capable of utilizing nitrate for phosphorus uptake under aerobic conditions partially contribute to EBPR. However, there exists no biological evidence that DNPAOs are actually predominant in the RAN reactor. This is mainly because all the previous studies have been conducted within the limitations of characterizing the total fraction of DNPAOs in a reactor through chemical analysis of the key biochemical transformation responsible for EBPR [22–24]. Although some studies have used molecular techniques such as PCR-DGGE, FISH and dot blot hybridization to detect PAOs capable of accumulating phosphorus under aerobic conditions, there exists no available information on the microbial community associated with the proportion of DNPAOs capable of utilizing only nitrate under aerobic or anoxic conditions.

3.2. Phosphorus uptake ability

In order to clarify the phosphorus uptake abilities of the sludges cultivated in RAN, RAA and RNN reactors, batch tests were conducted under different electron acceptor conditions. On the basis of the EBPR performance, it was expected that the sludges would show various phosphorus uptake abilities depending on the electron acceptor conditions, because it was suggested that PAOs were selectively predominated by the type of electron acceptor provided for EBPR.

Fig. 2 shows the correlation between the kind of electron acceptor used and the phosphorus uptake ability. The phosphorus uptake abilities of the sludges obtained from both RAN and RAA reactors exhibited no considerable differences in the cases with and without nitrate as electron acceptors under aerobic conditions. The amount of phosphorus uptake was 13.3 mg-P g^{-1}-MLSS (with nitrate) and 14.2 mg-P g^{-1}-MLSS (without nitrate) in the RAN reactor and 12.5 mg-P g^{-1}-MLSS

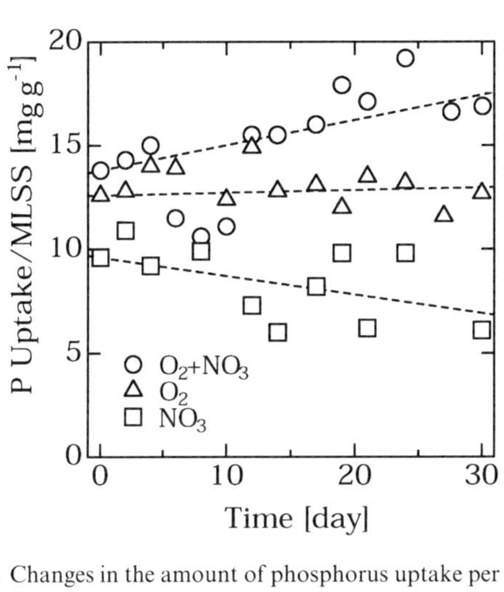

Fig. 1. Changes in the amount of phosphorus uptake per mixed liquor suspended solid (MLSS) under different electron acceptor conditions: (○), RAN; (△), RAA; and (□), RNN reactor.

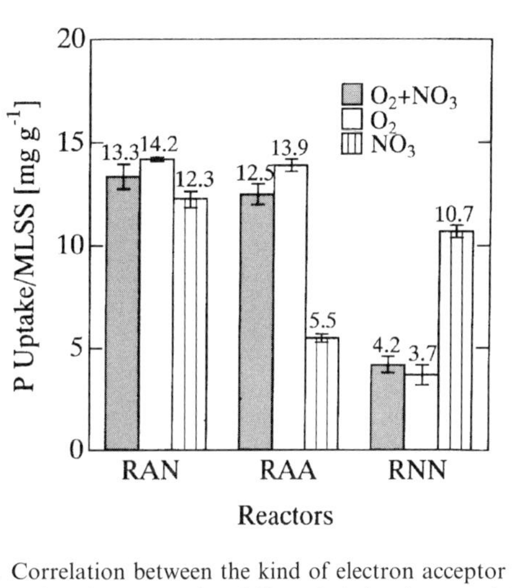

Fig. 2. Correlation between the kind of electron acceptor used and the phosphorus uptake ability of sludge harvested from each reactor operated under either one of the three different electron acceptor conditions.

(with nitrate) and 13.9 mg-P g^{-1}-MLSS (without nitrate) in the RAA reactor. However, the amount of phosphorus uptake per MLSS in the RAA reactor (5.5 mg-P g^{-1}-MLSS) was significantly low under anoxic conditions, while that in the RAN reactor (12.3 mg-P g^{-1}-MLSS) was similar to other electron acceptor conditions. The difference in phosphorus uptake ability between RAN and RAA reactors under anoxic conditions indicates that a proportion of DNPAOs that are able to take up phosphorus using only nitrate as electron acceptor are apparently present in the RAN reactor. However, comparison of the phosphorus uptake abilities in the presence of oxygen with and without nitrate reveals that DNPAOs do not play a major role in phosphorus uptake under aerobic conditions. In addition, in the RAA reactor, phosphorus uptake occurred even under anoxic conditions despite its being nitrification-inhibited during the operation period. This might be because a proportion of DNPAOs that were facultatively able to utilize oxygen or nitrate according to the electron acceptor conditions were present. This result agrees with the report of [25] that the periodical exposure of PAOs capable of anoxic phosphorus removal to oxygen did not affect phosphorus uptake activity under anoxic conditions.

On the other hand, the RNN reactor showed various phosphorus uptake abilities depending on the electron acceptor conditions. In contrast to RAN and RAA reactors, the phosphorus uptake ability in the RNN reactor under aerobic conditions when oxygen with and without nitrate was provided was significantly suppressed to one-half its ability under anoxic conditions, which resulted in 4.2 mg-P g^{-1}-MLSS (oxygen together with nitrate), 3.7 mg-P g^{-1}-MLSS (oxygen together without nitrate) and 10.7 mg-P g^{-1}-MLSS (only nitrate). This seems to be due to the fact that the RNN reactor contains a relatively low proportion of PAOs that are capable of utilizing only oxygen for phosphorus uptake, while it abundantly contains DNPAOs that are able to utilize only nitrate. Therefore, a proportion of DNPAOs has the facultative ability to utilize both oxygen and nitrate depending on the electron acceptor conditions, and the type of electron acceptor used for phosphorus uptake leads to a change in the microbial community structure of PAOs responsible for EBPR.

Accordingly, this finding suggests that the dominant PAOs consist of at least three types of dominant PAOs: the first one is facultatively able to utilize nitrate as well as oxygen depending on the electron acceptor conditions, the second one is capable of utilizing only nitrate, and the third one is capable of utilizing only oxygen for phosphorus uptake.

3.3. Analysis of DGGE pattern and phylogenetic affiliation

The PCR–DGGE method was adapted to obtain a better understanding with respect to the change in microbial community depending on the electron acceptor conditions, and to analyze the phylogenetic affiliation based on the PCR-amplified genes coding for 16S rDNA.

Fig. 3 shows DGGE profiles of 16S rDNA fragments of sludge samples obtained from each reactor that was

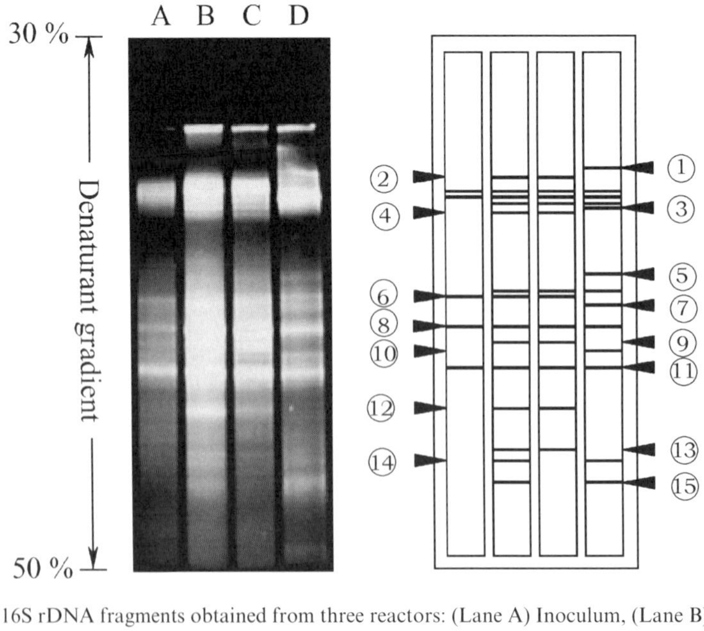

Fig. 3. DGGE profiles of 16S rDNA fragments obtained from three reactors: (Lane A) Inoculum, (Lane B) RAN, (Lane C) RAA and (Lane D) RNN reactors.

operated under different electron acceptor conditions. There existed many visible bands showing complexity and diversity of microbial ecology in the sludge samples. Based on the analysis of the DGGE patterns of each PCR-amplified 16S rDNA, lanes B and C showed similar band patterns, whereas two bands disappeared in lane C. Considering the results described in Fig. 2 in which the amount of phosphorus uptake per MLSS of sludge cultivated in the RAA reactor significantly decreased under anoxic conditions, it seems that the disappearance of bands 14 and 15 in lane C is associated with the decrease in the number of DNPAOs, because of the inhibition of nitrification under aerobic conditions.

On the other hand, a significant change in band patterns in lane D in relation to the absence of oxygen and the presence of nitrate was observed compared with

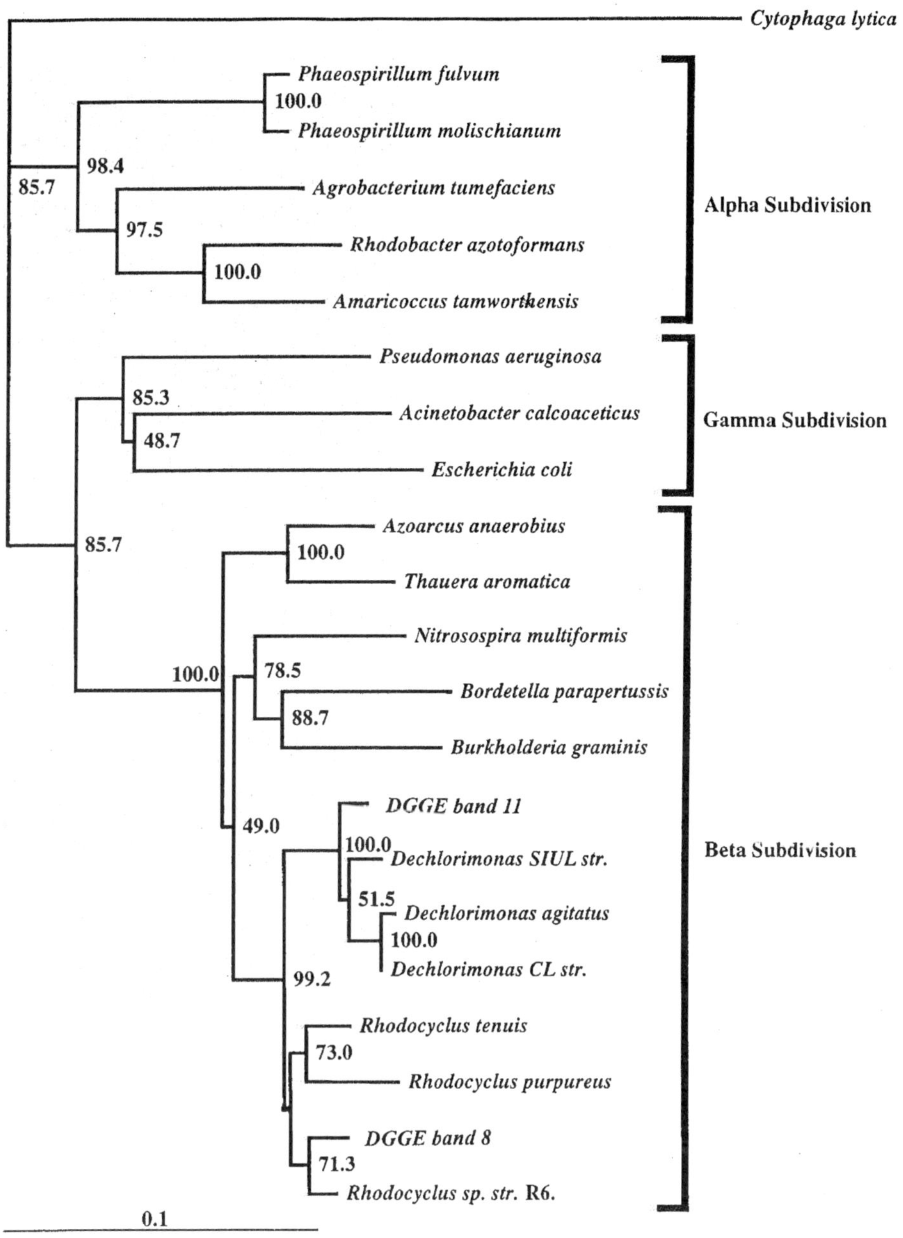

Fig. 4. Phylogenetic tree of 16S rDNA excised from DGGE profiles. Evolutionary distance and parsimonious analyses were carried out in CLUSTAL W employing 1000 bootstrap resamplings. The scale indicates 0.1 nucleotide substitution per nucleotide position. The tree was rooted with 16S rDNA sequence of *Cytophaga lytica* as an outgroup.

the other lanes: it showed the appearance of five new bands (1, 3, 5, 7 and 10) and the disappearance of six bands (2, 4, 6, 9, 12 and 13) that were commonly included in lanes B and C. This result suggests that the microbial community participating EBPR can be changed depending on the electron acceptor conditions. As discussed in Fig. 2, each reactor showed various abilities of phosphorus uptake per MLSS under different electron acceptor conditions, which suggested that the phosphorus uptake ability related to the microbial community of PAOs changed depending on the electron acceptor conditions. Therefore, the results suggest that some of the visible bands of DGGE patterns are associated with the change in the microbial community of PAOs. We initially expected that the microbial community of the RNN reactor would become less diverse, because the population of PAOs capable of utilizing only oxygen would decrease when nitrate alone was supplied as electron acceptor. However, in contrast to our expectation, lane D showed more complex band patterns than the other lanes. Thus, this result for the RNN reactor also seems to be related to the change in the microbial community of PAOs.

Based on the analysis of DGGE band patterns, it can be seen that bands 8 and 11 were commonly found in all the lanes. Since all the reactors despite being operated under different electron acceptor conditions showed the ability to take up phosphorus using both oxygen and nitrate, bands 8 and 11 might correspond to DNPAOs capable of utilizing both oxygen and nitrate as electron acceptors. Thus, bands 8 and 11 were excised from DGGE polyacrylamide gel, and then their sequences were analyzed. Furthermore, the obtainable sequences were compared with those available in GenBank, and then phylogenetic affiliation was determined.

Phylogenetic affiliation of the 16S rDNA of bands 8 and 11 is summarized in Fig. 4. It was found that bands 8 and 11 corresponded to related sequences of *Rhodocyclus* sp. (96% identity) and *Dechlorimonas* sp. (97% identity), respectively, both of which belong to the beta subclass of Proteobacteria. This finding, especially *Rhodocyclus* sp., is consistent with the results in which an important group of PAOs responsible for EBPR is closely related to *Rhodocyclus* sp. [9,12] and *Accumulibacter phosphatis* [13], whereas *Dechlorimonas* sp. remains uncharacterized.

3.4. In situ detection of dominant species using FISH analysis

As a result of PCR–DGGE and DNA sequencing, it was suggested that *Rhodocyclus* sp., one of the candidates for PAOs, is commonly present in all the reactors performing EBPR under different electron acceptor conditions. Nevertheless, DGGE fingerprints obtained from electrophoresis did not give any information on the extent of its predominance in the microbial community. Thus, FISH analysis was conducted to detect *Rhodocyclus* sp. and *Dechlorimonas* sp., because in situ hybridization with 16S rRNA-targeted oligonucleotide probes enables the identification and monitoring of the specific strains in the microbial community.

Figs. 5(A) and (B) show FISH images of *Rhodocyclus* sp., specific for oligonucleotide probes in the sludge sample obtained from the RNN reactor. *Rhodocyclus* sp.

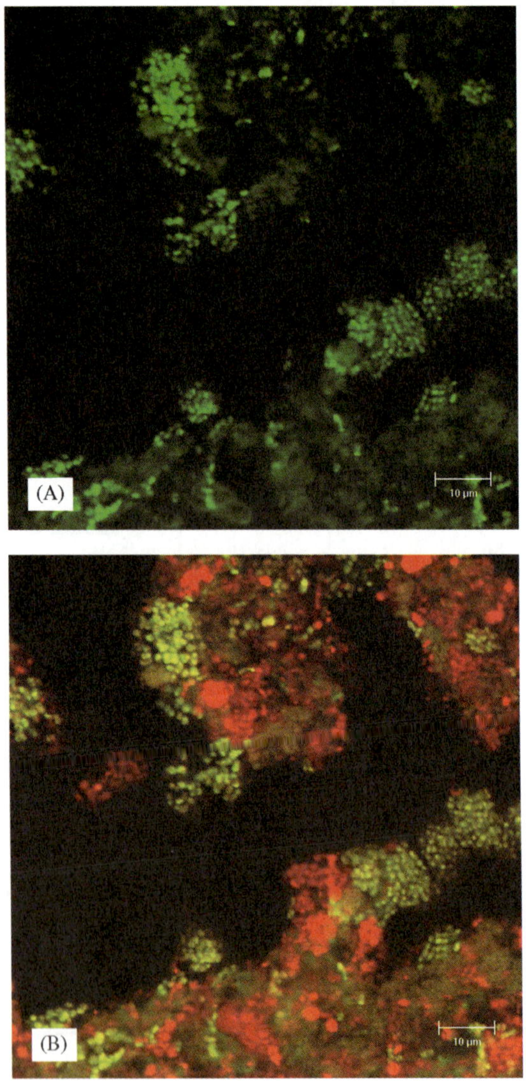

Fig. 5. FISH images of *Rhodocyclus* sp. specific for oligonucleotide probes in the sludge sample obtained from the RNN reactor. Image of the sludge sample (A) single-hybridized with PAO846 labeled with FITC (green) and (B) dual-hybridized with both FITC (green)-labeled PAO846 probe and CY3 (red)-labeled EUB388 probe. The yellow color represents *Rhodocyclus* sp., and the red color indicates other bacteria, bar = 10 µm.

in situ dual-hybridized with EUB388 (red) and PAO846 (green) probes showed clear yellow signals, which demonstrated that *Rhodocyclus* sp. occupies a considerable proportion of the RNN reactor. Moreover, FISH analysis of the other reactors also demonstrated that *Rhodocyclus* sp. is abundantly present, similar to the RNN reactor (data not shown). In addition, morphological observation of FISH images revealed that *Rhodocyclus* sp. in situ hybridized with PAO846 probe (green) forms clusters resembling bunches of grapes, as can be seen in Fig. 5(A), and is morphologically similar to that in the other reactors. This result suggests that *Rhodocyclus*-like organism whose partial 16S rDNA sequence was isolated from DGGE polyacrylamide gel is capable of facultatively utilizing both oxygen and nitrate as electron acceptors for phosphorus uptake, and thus plays a major role in EBPR under different electron acceptor conditions.

However, hybridization using probe DMC 488 showed that *Dechlorimonas* sp. did not exist abundantly in the sludge samples obtained from all the reactors (data not shown), which suggested that *Dechlorimonas* sp. was not dominant population in all the reactors, while it was detected by DGGE. This is probably because the 16S rDNA of *Dechlorimonas* sp. extracted from the sludge samples was exaggeratedly amplified by PCR using universal primers despite its small proportion of microbial community.

Accordingly, FISH analysis demonstrates that *Rhodocyclus* sp. is present in all the reactors fed with acetate as the carbon source even though the electron acceptor conditions differ from each other.

4. Conclusions

In this paper, we presented the characteristics and the microbial diversity of DNPAOs that are capable of conducting anoxic phosphorus uptake using nitrate as electron acceptor. From this study, the following results were obtained:

1. The sludges obtained from each reactor showed various abilities of phosphorus uptake per MLSS under different electron acceptor conditions, which revealed that PAOs responsible for EBPR consisted of at least three populations including DNPAOs, and the microbial community structure was changed depending on the electron acceptor conditions.
2. The sludge cultivated with oxygen together with nitrate showed a drastic increase in the amount of phosphorus uptake under anoxic conditions, which suggested that a proportion of DNPAOs capable of utilizing nitrate under aerobic conditions were present.
3. The results of DGGE of PCR-amplified 16S rDNA fragments showed that the change in the microbial community that took place depending on the type of electron acceptor, and the determination of phylogenetic affiliation demonstrated that the bacteria commonly contained in the three reactors were *Rhodocyclus* sp.that belonged to the beta subclass of Proteobacteria.
4. FISH analysis demonstrated that *Rhodocyclus* sp. occupied a considerable proportion in all the reactors even though the electron acceptor conditions differed from each other.

References

[1] Bortone G, Marsili Libelli S, Tilche A, Wanner J. Anoxic phosphate uptake in the DEPHANOX process. Water Sci Technol 1999;40(4–5):177–85.

[2] Kerrn-Jespersen JP, Henze M, Strube R. Biological phosphorus release and uptake under alternating anaerobic and anoxic conditions in a fixed-film reactor. Water Res 1994;28(5):1253–5.

[3] Kuba T, Smolder G, van Loosdrecht MCM, Heijnen JJ. Biological phosphorus removal from wastewater by anaerobic-anoxic sequencing batch reactor. Water Sci Technol 1993;27(5–6):241–52.

[4] Šorm R, Bortone G, Saltareli R, Jenèek P, Wanner J, Tilche A. Phosphate uptake under anoxic condition in nutrient removal activated sludge system. Water Res 1996;30(7):1573–84.

[5] Šorm R, Bortone G, Wanner J, Tilche A. Behaviour of activated sludge from a system with anoxic phosphate uptake. Water Sci Technol 1996;37(4–5):563–6.

[6] Wanner J, Cech JS, Kos M. New process design for biological nutrient removal. Water Sci Technol 1992;25 (4–5):445–8.

[7] Kuba T, van Loosdrecht MCM, Heijnen JJ. Phosphorus and nitrogen removal with minimal COD requirement by integration of denitrification in a two-sludge system. Water Res 1996;30(7):1702–10.

[8] Henze M, Gujer W, Mino T, Matsuo T, Wentzel MC, Marais GvR, Van Loosdrecht MCM. Activated sludge model no. 2d, ASM2D. Water Sci Technol 1999;39(1): 165–82.

[9] Bond PL, Hugenholtz P, Keller J, Blackall LL. Bacterial community structures of phosphate-removing and non-phosphate-removing activated sludges from sequencing batch reactors. Appl Environ Microbiol 1995;61(5): 1910–6.

[10] Bond PL, Erhart R, Wagner M, Keller J, Blackall LL. Identification of some of the major groups of bacteria in efficient and nonefficient biological phosphorus removal activated sludge systems. Appl Environ Microbiol 1999;65(9):4077–84.

[11] Bond PL, Keller J, Blackall LL. Bio-p and non bio-p bacteria identification by a novel microbial approach. Water Sci Technol 1999;39(6):13–20.

[12] Crocetti GR, Hygenholtz P, Bond PL, Schuler A, Keller J, Jenkins D, Blackall LL. Identification of polyphosphate-accumulating organisms and design of 16S rRNA-directed

probes for their detection and quantitation. Appl Environ Microbiol 2000;66(3):1175–82.

[13] Hesselmann RPX, Werlen C, Hahn D, van der Meer JR, Zehnder AJB. Enrichment, phylogenetic analysis and detection of a bacterium that performs enhanced biological phosphate removal in activated sludge. Systems Appl Microbiol 1999;22:454–65.

[14] Muyzer G, De Waal EC, Uitterlinden AG. Profiling of complex microbial populations by denaturing gradient gel electrophoresis analysis of polymerase chain reaction-amplified genes coding for 16S rRNA. Appl Environ Microbiol 1993;59(3):695–700.

[15] Muyzer G, Ramsing NB. Molecular methods to study the organization of microbial communities. Water Sci Technol 1995;32(8):1–9.

[16] Ishii K, Nakagawa T, Fukui M. Microbes Environ 2000;15(1):59–73 (in Japanese).

[17] APHA, AWWA, WEF. Standard methods for the examination of water and wastewater. 18th ed. American Public Health Association, Washington, DC, 1992.

[18] Amann RI, Krumholz L, Stahl DA. Fluorescent-oligonucleotide probing of whole cells for determinative, phylogenetic, and environmental studies in microbiology. J Bacteriol 1990;172(2):762–70.

[19] Amann RI. In situ identification of micro-organisms by whole cell hybridization with rRNA-targeted nucleic acid probes. In: Akkermans ADL, Elsas JD, de Bruij FJ, editors. Molecular microbial ecology manual, vol. 3. 3. 6. London: Kluwer, 1995. p. 1–15.

[20] Kuba T, Murnleitner E, van Loosdrecht MCM, Heijnen JJ. A metabolic model for biological phosphorus removal by denitrifying organisms. Biotech Bioeng 1996;52:685–95.

[21] Murnleitner E, Kuba T, van Loosdrecht MCM, Heijnen JJ. An integrated metabolic model for the aerobic and denitrifying biological phosphorus removal. Biotech Bioeng 1997;54(5):434–50.

[22] Meinhold J, Filipe CDM, Daigger GT, Isaacs S. Characterization of the denitrifying fraction of phosphate accumulating organisms in biological phosphate removal. Water Sci Technol 1999;39(1):31–42.

[23] Meinhold J, Arnold E, Isaacs S. Effect of nitrite on anoxic phosphate uptake in biological phosphorus removal activated sludge. Water Res 1999;33(8):1871–83.

[24] Wachtmeister A, Kuba T, van Loosdrecht MCM, Heijnen JJ. A sludge characterization assay for aerobic and denitrifying phosphorus removing sludge. Water Res 1997;31(3):471–8.

[25] Kuba T, van Loosdrecht MCM, Heijnen JJ. Effect of cyclic oxygen exposure on the activity of denitrifying phosphorus removing bacteria. Water Sci Technol 1996;34(1–2):33–40.

PERGAMON

Water Research 36 (2002) 413–420

WATER RESEARCH

www.elsevier.com/locate/watres

Performance of a bioreactor with submerged membranes for aerobic treatment of municipal waste water

S. Rosenberger[a], U. Krüger[a], R. Witzig[b], W. Manz[b], U. Szewzyk[b], M. Kraume[a,*]

[a] *Department of Chemical Engineering, Technical University of Berlin, Strasse des 17. Juni 136, D-MA 5-7, Secretariat 10623 Berlin, Germany*

[b] *Department of Microbial Ecology, Technical University of Berlin, Franklinstrasse 29, D-OE 5, Secretariat 10587 Berlin, Germany*

Received 20 March 2000; received in revised form 29 December 2000; accepted 1 February 2001

Abstract

Aerobic treatment of municipal waste water in a membrane bioreactor was studied for 535 d. Apart from sampling, sludge was retained completely by a submerged hollow fibre membrane with a pore-size of 0.2 μm. The pilot plant comprised an anoxic zone to enable denitrification. The maximum liquid hold-up of the plant was $3.9\,m^3$. In this study the reactor performance and the stability of the process and the membrane capacity were investigated. A stable flux of $18\,l\,m^{-2}\,h^{-1}$ could be realised with a mean transmembrane pressure difference of 0.3 bar with air-bubbling and backflushing the membrane and cleaning it in place every two months for one or two hours. For about 140 d, a flux of $27\,l\,m^{-2}\,h^{-1}$ was achieved, but cleaning became necessary more often. The hydraulic retention time (HRT) varied between 10.4 and 15.6 h. Accordingly the volumetric loading rate was between 1.1 and $1.7\,kg\,COD\,m^{-3}\,d^{-1}$. No inoculum was used. The mixed liquor suspended solids (MLSS) concentration gradually increased to $18–20\,g\,MLSS\,l^{-1}$. The feed to microorganism (F/M) ratio varied according to the operation conditions but decreased against a value of $0.07\,kg\,COD\,kg^{-1}\,MLSS\,d^{-1}$. Treatment performance was very stable and on a high level. The COD was reduced by 95%. Nitrification was complete and up to 82% of the total nitrogen could be denitrified.

Keywords: Activated sludge; Complete sludge retention; Membrane bioreactor; Sludge production; Submerged membrane

1. Introduction

In municipal waste water treatment demands on effluent quality still increase. Not only carbon and nutrients but also bacteria and viruses have to be eliminated to a great extent according to the EEC Council directive [1]. Furthermore, the disposal of sludge containing too much heavy metals becomes increasingly difficult and costly as the sludge can no longer be deposited on agricultural land.

In recent years, new membranes specially developed for the use in waste water treatment have made membrane bioreactors become a promising alternative to the well-known aerobic processes. Ultra- or microfiltration membranes replace conventional sedimentation for the separation of the treated water from the sludge. The use of submerged membranes has reduced the power consumption of membrane bioreactors significantly and hence increased the potential for the application of membranes in waste water treatment. Moreover, ultra- and microfiltration membranes with a pore-size of 0.2 μm or less not only retain bacteria but also viruses virtually completely [2–6].

The complete retention of sludge allows operation at much higher biomass concentrations. The higher the concentration the lower the F/M ratio becomes, with the effect that the microorganisms utilise a growing portion of the carbon content of the feed for maintenance

*Corresponding author. Tel.: +49-30-314-23701; fax: +49-30-314-21134.

E-mail address: vtsekr@ivtfgl.tu-berlin.de (M. Kraume).

0043-1354/02/$ - see front matter
PII: S0043-1354(01)00223-8

Nomenclature

$B_{V,COD}$	volumetric load, kg COD $m^{-3} d^{-1}$
COD	chemical oxygen demand, mg l^{-1}
F/M	feed to microorganism ratio, kg COD kg^{-1} MLSS d^{-1}
HRT	hydraulic retention time, h
MLSS	mixed liquor suspended solids (sludge concentration), g MLSS l^{-1}
MLVSS	mixed liquor volatile suspended solids, g MLVSS l^{-1}
NH_4-N	ammonium nitrogen concentration, mg l^{-1}
NO_2-N	nitrite nitrogen concentration, mg l^{-1}
NO_3-N	nitrate nitrogen concentration, mg l^{-1}
N_t	total nitrogen concentration, mg l^{-1}
PO_4-P	phosphate phosphorus concentration, mg l^{-1}
Q_0	influent flow rate, $m^3 h^{-1}$(equivalent to the throughput)
Q_{DN}	flow rate of the recirculation to the denitrification compartment, $m^3 h^{-1}$
Q_F	flow rate of the recirculation to the filtration compartment, $m^3 h^{-1}$
V_{DN}	volume of the denitrification compartment, m^3 (0.7 m^3)
V_F	volume of the filtration compartment, m^3 (0.7 m^3)
V_N	volume of the nitrification compartment, m^3 (max. 2.5 m^3)

purposes and consequently less for growth. When the sludge load becomes low enough, no or almost no excess sludge is produced any more [7–10]. Stephenson et al. [6] presented literature data on sludge production and concluded that extremely low sludge production is possible for low F/M ratios and high sludge ages. In addition, high sludge ages allow specialised microorganisms, as for example nitrifiers, to establish themselves in the system. They are no longer washed out, and hence stable nitrification can be observed as long as the oxygen concentration is high enough [2,5,7,9,11]. When anoxic zones exist, denitrification occurs and seems even to be stimulated by high biomass concentrations [12]. Another advantage of the use of membranes is the indifference to hydraulic fluctuations. Sludge bulking is no longer a problem. Because of higher volumetric loading rates, smaller reactor volumes can be realised, and secondary clarification and possibly parts of the pre-treatment can be omitted.

Earlier studies have already shown that membrane bioreactors can be operated at much higher biomass concentrations than conventional treatment plants for long periods and with complete sludge retention. Treatment performances were generally better, and deterioration of the performance was not observed [2,7,9,11,13,14]. The microbiological processes which are at the basis of the biological treatment are not yet understood very well. However, the understanding is very important for the design, the control, and the optimisation of the process.

The aim of this study was to investigate the long-term treatment performance of a membrane bioreactor with complete sludge retention for municipal waste water taking the microbiology into consideration. The influence of the biomass concentration and the hydraulic retention time were examined as well as the composition and the structure of the biocoenosis and the activity of the microorganisms [21]. To follow the development of the treatment performance, biomass evolution and carbon, nitrogen, and phosphorus removal were studied. Reactor volume and flow rates were varied to give different hydraulic retention times.

2. Materials and methods

2.1. Plant configuration

The pilot plant was constructed by Zenon GmbH, Düsseldorf and installed at the German Federal Environmental Agency in Berlin. The configuration is shown in Fig. 1. The pilot plant consisted of three compartments with a total volume of 3.9 m^3. The nitrification compartment (2.5 m^3) and the filtration compartment (0.7 m^3) were aerated, the denitrification compartment (0.7 m^3) was stirred. Thus all compartments can be considered to be completely mixed. The sludge was retained by a submerged microfiltration membrane module.

2.2. Membrane

The ZeeWeed^{TM} module ZW150 (Zenon) consisted of neutral hydrophilic hollow fibre membranes with a pore-size of 0.2 µm. The total surface area was 13.9 m^2. The treated water was sucked through the membrane due to a partial vacuum inside the hollow fibres created by a pump. The membrane module was aerated to keep the hollow fibres moving and create a flow along the membrane surface, so that membrane fouling was reduced to a minimum. Every 10 min the flow through the membranes was reversed for 35 s to backflush the membranes. When the transmembrane pressure difference exceeded 0.5 bar, the membranes were cleaned in place. For this purpose the hollow fibres were rinsed inside with citric acid, caustic soda, or sodium hypochloride which was also used for disinfection.

2.3. Waste water

In this research, presettled municipal waste water from the neighbouring sewage works 'Berlin-Marienfelde' was treated. The waste water was screened (2 mm) before entering the denitrification compartment. The average feed concentrations are given in Table 1. The waste water contained approximately 4% non-degradable COD [15] as 'Berlin-Marienfelde' is equipped with a thermal sludge treatment and the waste water is recirculated to the inlet. The COD/BOD_5 ratio was approx. 2.4.

2.4. Operating conditions

In order to gain insight into the development of the biocoenosis in membrane bioreactors, the pilot plant was started up without addition of activated sludge, e.g. of a conventional sewage works as would be the case for full-scale reactors. Thus only the biomass introduced with the waste water was used as inoculation. During the whole period of the study (535 d), no sludge was removed from the plant intentionally except for sampling. However, there were some incidents with accidental sludge loss (see Section 3). The oxygen concentration was always 2 mg l^{-1} apart from some short technical faults. The permeate flow rate was kept at a desired value, whereas the feed was controlled by a level detector to keep a certain level in the reactor. Because of changes in flow rates and the volume of the nitrification compartment, four periods with different hydraulic retention times (HRT) were distinguished according to Table 2. During the first period the influent flow rate Q_0 was 0.25 m^3 h^{-1} and the volumetric loading rate ($B_{V,COD}$) 1.2 kg COD m^{-3} d^{-1}. HRT was 15.6 h. In the second period the flow rate was increased by 50% to reduce the HRT to 10.4 h ($B_{V,COD}$ = 1.8 kg COD m^{-3} d^{-1}). The periods 2a and b differ in the flow rate of the recirculation to the denitrification compartment. In the third period the flow rate was reduced back to 0.25 m^3 h^{-1} but the nitrification volume was reduced at the same time to keep the HRT constant at 10.4 h.

To see whether a similar state of the biomass as in the first period would be reached, the volume was increased again in the fourth period to give conditions almost similar to those in the first period.

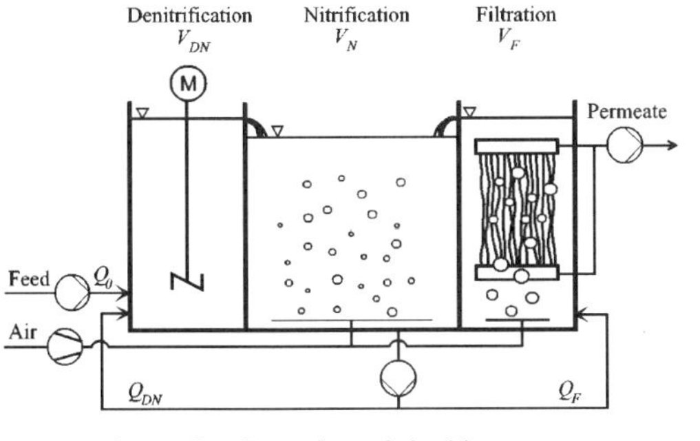

Fig. 1. Configuration of the bioreactor.

2.5. Analysis

The chemical oxygen demand (COD), concentrations of ammonium, nitrite, nitrate nitrogenand total nitrogen (NH_4-N, NO_2-N, NO_3-N, and N_t) as well as phosphate phosphorus (PO_4-P) in feed and effluent were determined regularly with standard vial tests of Dr. Lange: LCK 314 (15–150 mg l^{-1} COD), LCK 514 (100–2000 mg l^{-1} COD), LCK 304 (0.015–2.0 mg l^{-1} NH_4-N), LCK 303 (2–47 mg l^{-1} NH_4-N), LCK 341 (0.015–0.6 mg l^{-1} NO_2-N), LCK 342 (0.6–6.0 mg l^{-1} NO_2-N), LCK 339 (0.23–13.5 mg l^{-1} NO_3-N), LCK 238 (1–40 mg l^{-1} N_t), LDK 338 (20–100 mg l^{-1} N_t), and LCK 350 (2–20 mg l^{-1} PO_4-P). For the sludge, the concentrations of mixed liquor suspended solids (MLSS) and mixed liquor volatile suspended solids (MLVSS) were determined according to the German Standards for the Examination of Water and Waste Water (DIN 38 414).

Table 1
Mean concentrations of the feed

	COD (mg l^{-1})	NH_4-N (mg l^{-1})	NO_2-N (mg l^{-1})	NO_3-N (mg l^{-1})	N_t (mg l^{-1})	PO_4-P (mg l^{-1})	MLSS (g l^{-1})	MLVSS (%)
Min. conc.	299	21.2	—	0.08	31.8	4.9		
Mean conc.	786	49.4	0.04	0.75	65.8	11.8	1.1	32
Max. conc.	4294	78.2	0.13	3.82	202.4	106.2		

Table 2
Operating conditions

Period	Days	V_N (m^3)	V_{DN} (m^3)	V_F (m^3)	Q_0 (m^{3}h^{-1})	Q_{DN} (m^{3}h^{-1})	Q_F (m^{3}h^{-1})	HRT (h)	$\overline{B_{V,COD}}$ (kg COD m^{-3}d^{-1})
1	0–110	2.5	0.7	0.7	0.25	1	5	15.6	1.1
2a	111–170	2.5	0.7	0.7	0.375	1.5	5	10.4	1.7
2b	171–254	2.5	0.7	0.7	0.375	1.75	5	10.4	1.7
3	258–314	1.2	0.7	0.7	0.25	1.25	5	10.4	1.6
4	315–535	2.1	0.7	0.7	0.25	1.25	5	14	1.2

3. Results and discussion

The pilot plant has been running for 535 d without any severe problems. However, there were some technical failures, which are mentioned below. The results are classified in four periods according to the different operating conditions (see Table 2).

3.1. Operating conditions

The pH was always stable between 7 and 7.5. The reactor temperature was significantly higher (20–27°C) than the temperature of the influent (12–20°C). It has to be noted that the pilot plant was erected in a laboratory which was held at temperatures between 15°C in the winter and 25°C in the summer. An energy balance showed that about 3.3°C of the temperature rise resulted from microbial activity, i.e. the total oxidation of the carbon and nitrogen supplied. The aeration did not have a significant effect on the temperature, neither by the power input of the blower nor by the evaporation of water which is taken out by the air.

3.2. Membrane performance

Before start-up with waste water, the pilot plant was tested with tap water for five weeks with a flux of 22 l m^{-2}h^{-1}. During the first period, the flux was set to 18 l m^{-2}h^{-1}. Immediately after start-up with waste water, the pressure difference increased to 0.33 bar (Fig. 2). The abrupt decline is due to the deposition of, e.g. slimy substances and lipids. Experiments with biofilms showed that these substances adhere very quickly to surfaces (conditioning phase) whereas bacterial growth of microorganisms in biofilms only starts after several hours or even days [16]. The pressure difference then stabilised between 0.18 and 0.26 bar as the sludge concentration increased.

In the second period, an increased flux from 25 to 27 l m^{-2}h^{-1} was realised. To achieve the flux, the necessary transmembrane pressure difference increased from 0.3 to 0.55 bar. The membrane had to be cleaned more often, as the transmembrane pressure difference exceeded 0.5 bar. In general, the performance was less stable than in the first period. This probably resulted in a

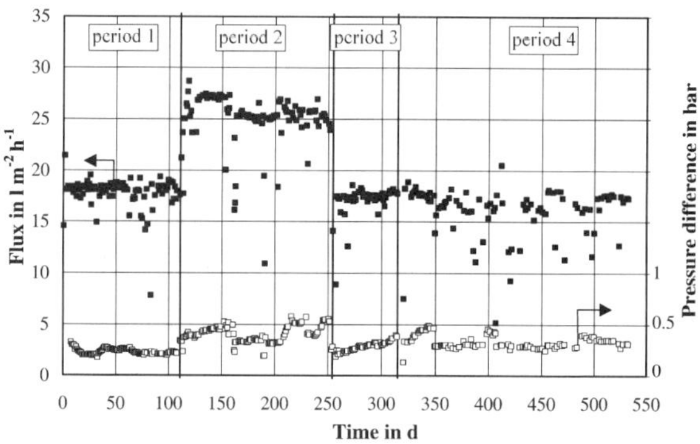

Fig. 2. Flux and pressure difference during the different operation periods.

layer on the membrane which was more compact and reduced the permeability. Whether a flux of 27 l m^{-2}h^{-1} can be stabilised with intensified backflushing or bubbling should be the subject of another study.

In the third period, the flux was reduced to 18 l m^{-2}h^{-1} again. By cleaning the membrane with sodium hypochloride the pressure difference was reduced to 0.18 bar but increased rapidly to 0.36 bar afterwards.

During the fourth period, a stable flux of 18 l m^{-2}h^{-1} with a medium pressure difference of 0.3 bar could be achieved. Although the biomass concentration increased to about 18–20 g MLSS l^{-1}, a significant decline in permeability was not observed.

3.3. Development of the biomass concentration

The development of the biomass concentration was followed by determining the MLSS and MLVSS. In the first period, the biomass developed from the microorganisms coming in with the waste water as the reactor had not been inoculated. As given in Fig. 3 the MLSS increased rapidly at the beginning and more slowly with increasing sludge concentration. Finally the sludge concentration stabilised at about 9 g l^{-1}. As the feed concentrations varied over a wide range, the sludge concentration was not absolutely constant but fluctuated around a mean value. The MLVSS stabilised at

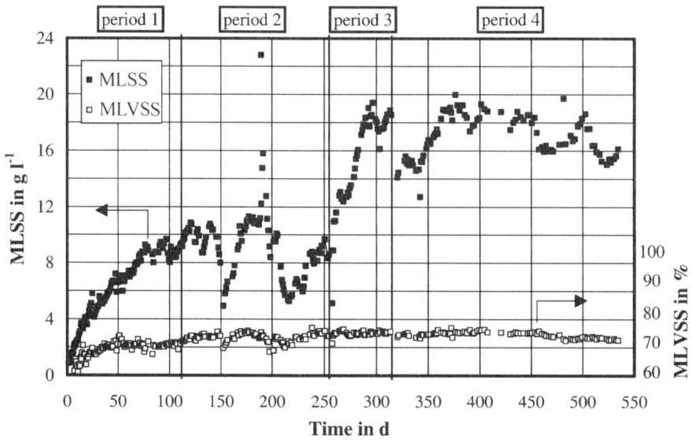

Fig. 3. Development of the sludge concentration and the volatile suspended solids in the reactor.

70%, which is a common value in municipal waste water treatment.

In the second period, the HRT was reduced by increasing the volumetric loading rate to approximately 1.7 kg COD $m^{-3} d^{-1}$. This resulted in a growth of biomass. On day 152 sludge was lost because a level detector had failed. Thereafter the sludge concentration went up to approximately 11 g l^{-1}. On days 189 and 190 return sludge of the sewage works 'Marienfelde' was accidentally pumped into the pilot plant, resulting in a sharp increase of the sludge concentration to 22 g l^{-1}. During the following days, the sludge concentration decreased to 5 g MLSS l^{-1}. Afterwards the system recovered. On day 257, the recirculation to the filtration compartment was blocked so that effluent and feed stopped, leading to a decrease in MLSS concentration (5 g l^{-1}).

In the third period, the permeate flow rate was reduced by lowering the volume of the nitrification compartment. As the total amount of biomass was not affected, the sludge concentration increased up to values of 20 g l^{-1}.

In the fourth period, the total reactor volume was raised to 3.5 m^3 to re-establish conditions close to those of the first period. By this the reversibility of the process was investigated. Caused by the dilution, the sludge concentration first fell to 14 g l^{-1} but increased to the former value of 19 g l^{-1} again. During the following 150 d, a very slow decrease down to 16 g l^{-1} occurred. Obviously, the sludge concentration in mixed cultures is not only dependent on the volumetric loading rate. Under comparable conditions in the first period the biomass concentration had attained values around 9 g MLSS l^{-1}. Bouillot et al. [17] and Low and Chase [10] had shown for pure cultures that at constant breeding rates (sludge ages) the final biomass concentration depends only on the substrate mass flow.

When a constant biomass concentration was reached, the F/M ratio decreased against a value of 0.07–0.1 kg COD kg^{-1} MLSS d^{-1} as shown in Fig. 4. This value ranges between those found by Yamamoto [7], Wenger-Oehn and Braun [18] and Muller et al. [9]. Yamamoto obtained 0.1 kg COD kg^{-1} MLSS d^{-1} for synthetic waste water and Wenger-Oehn and Braun obtained 0.17 kg COD kg^{-1} MLSS d^{-1} for leachate. Muller, however, who worked with domestic waste water, observed only 0.021 kg COD kg^{-1} MLSS d^{-1}. This phenomenon of various F/M ratios may be due to different feed compositions and microbial populations. It is possible that there are shifts in the population which are induced by changes in the operating conditions.

The net sludge production for F/M ratios as low as 0.07 kg COD kg^{-1} MLSS d^{-1} decreased towards zero as shown in Fig. 5. Therefore, constant MLSS concentrations as shown in Fig. 3 can be linked to low F/M ratios. This is in accordance with the maintenance concept introduced by Pirt [19] which demands that biomass growth slows down with decreasing nutrient supply. Low F/M ratios equal little substrate per unit biomass, which leads to competition among the microorganisms and results in reduction of the net sludge production.

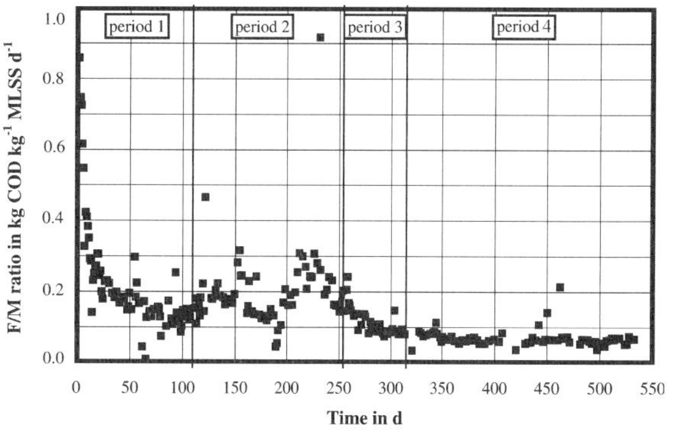

Fig. 4. Development of the F/M ratio with complete sludge retention.

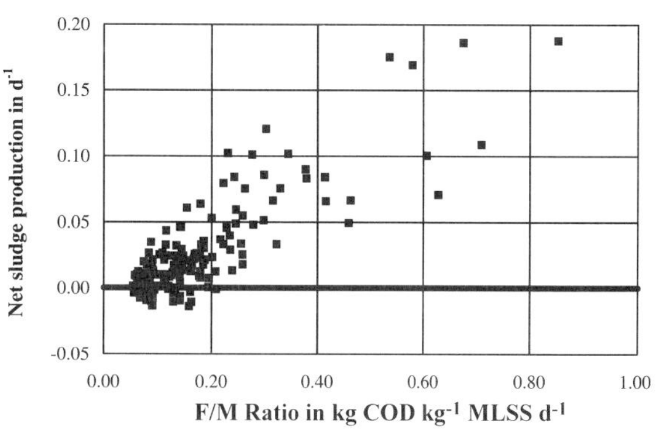

Fig. 5. Development of the net sludge production.

When no excess sludge is drawn from the reactor, accumulation of inorganic compounds must be expected. During this research, a stable value of about 75% was found in the activated sludge samples regularly drawn from the nitrification compartment. When emptying the reactor after the operation period, a dense sludge up to the height of aerators was found at the bottom of the reactor, which consisted about 60% of inorganic substances. Therefore, it was assumed that most inorganic substances settled on the bottom of the reactor in dead zones and hence could not be found among the suspended solids. Günder and Krauth [20] showed that the excess sludge production is not only a function of sludge load, but also of the amount of influent particulate and non biodegradable matter. The lowest sludge production rate they found for municipal waste water was 0.3 kg MLSS kg^{-1} BOD_5.

Another problem may arise when the viscosity of the sludge becomes too high. On day 295, with a MLSS of 19 $g\,l^{-1}$, the viscosity was measured in a rotating viscosimeter. At a shear rate of 542 s^{-1} a value of 9.47 mPa s was obtained, at 1.25 s^{-1} even 483 mPa s. The sludge showed obvious pseudoplastic properties.

3.4. Degradation performance

The degradation performance of the pilot plant was followed by determining the removal of carbon, nitrogen and phosphorus.

Right from the start, where the MLSS was still below 3 $g\,l^{-1}$, the COD concentration of the effluent did not exceed 100 $mg\,l^{-1}$ (Fig. 6). The degradation rate was already 90%. With increasing sludge concentration, the reduction reached 95% and the effluent concentration declined to around 35 $mg\,l^{-1}$. Generally the effluent concentration was remarkably stable. Even shock loads (e.g. an influent COD of 4000 $mg\,l^{-1}$ on days 189 and 190) did not affect the effluent concentration significantly. No influence due to the different hydraulic retention times was observed.

The reduction of 95% is very high compared with the reduction in conventional treatment plants (75–85% at low F/M ratios), in particular, as the waste water contained 4% non-degradable COD. This can be attributed firstly to the complete retention of all particulate COD by the membrane. Secondly the biomass can adapt to the waste water without being restricted to fast growing and floc-forming microorganisms as wash-out is avoided. Under these conditions specialised microorganisms can establish which are able to remove slowly degradable components. High removal efficiencies between 90 and 98% have been reported by various authors and seem to be one of the major advantages of membrane bioreactors [6].

The mean total nitrogen concentration in the influent was 65 $mg\,l^{-1}$. It consisted mainly of ammonium nitrogen (80%) and only 1% of nitrate nitrogen (Table 1). Nitrite nitrogen in the feed was negligible. Nitrogen can be removed only by assimilation into biomass or by nitrification-denitrification. In this case, the possible denitrification is dependent on the recirculation ratio through the denitrification compartment Q_{DN}/Q_0. During times of constant sludge concentration, the removal of nitrogen has to be completely attributed to nitrification–denitrification.

During start-up, nitrogen was only assimilated and most of the nitrogen went into the effluent. Ten days later, the ammonium nitrogen concentration declined rapidly to values below 0.1 $mg\,l^{-1}$ (Fig. 7). As nitrifying bacteria have longer generation times, they need more time to establish and reach sufficient concentrations to nitrify the ammonium. It took another ten days until nitrite was completely converted to nitrate. Denitrification started as soon as nitrite appeared but did not work properly until the beginning of the second period.

From the beginning of the second period, the nitrification/denitrification performance was very stable (Fig. 8). When the recirculation ratio Q_{DN}/Q_0 was set to four, up to a maximum of 4/5 (80%) of the nitrogen was eliminated. On day 170, Q_{DN}/Q_0 was set to five, so that up to 5/6 (83%) of the nitrogen were eliminated. In

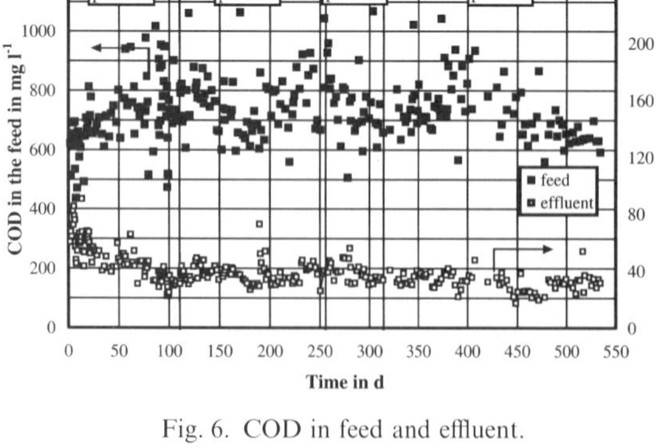

Fig. 6. COD in feed and effluent.

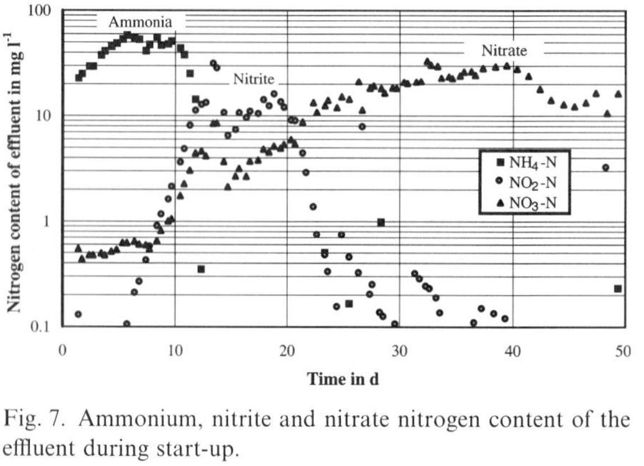

Fig. 7. Ammonium, nitrite and nitrate nitrogen content of the effluent during start-up.

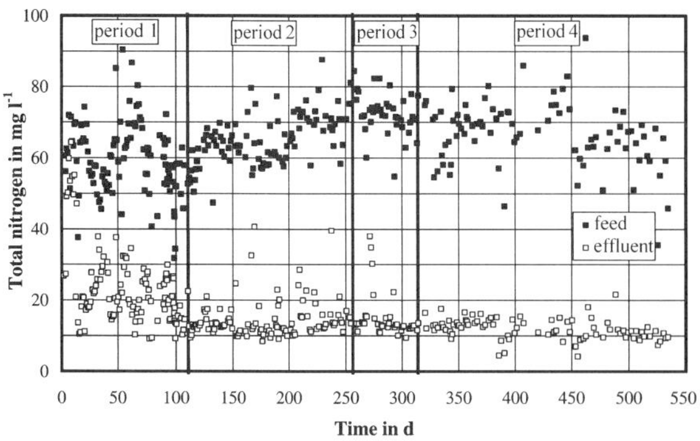

Fig. 8. Total nitrogen in feed and effluent.

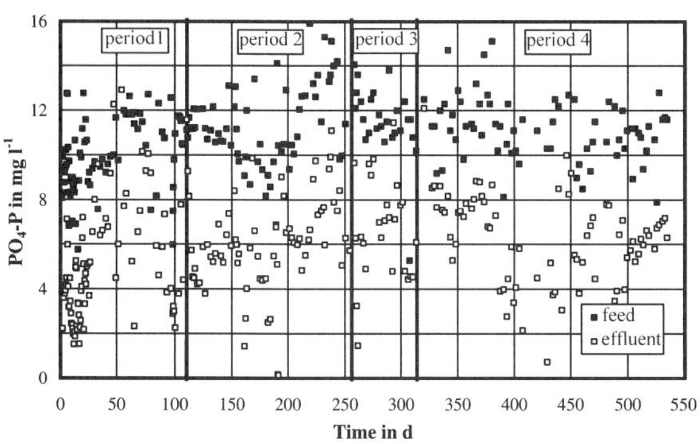

Fig. 9. Phosphate phosphorus in feed and effluent.

total, the mean concentration of nitrogen in the effluent, which was mostly nitrate nitrogen, was about $13\,mg\,l^{-1}$ and very stable.

Phosphate can be eliminated by two different mechanisms: assimilation and luxury uptake. Assimilation will only take place during the growth phases, luxury uptake only when the waste water passes through aerobic and anaerobic zones alternately. Both elimination mechanisms are bound to stop, precisely when the microorganisms do not grow any more and are saturated.

Surprisingly there has always been an elimination of around 40% during this investigation as shown in Fig. 9. A general mass balance shows that the total eliminated phosphate phosphorus up to the day 380 would make up about 20% of the total biomass contained in the reactor. This seems to be too much to be due only to luxury uptake of phosphate. It was assumed, that phosphate precipitated and together with inorganic substances has settled in dead zones on the bottom of the reactor and hence could not be found by effluent analysis.

4. Conclusion

Apart from some incidents the operation of the plant was stable throughout the total time of 535 d. Neither the concentrated biomass nor the membrane capacity caused any significant problems. A stable flux of $18\,l\,m^{-2}\,h^{-1}$ could be realised with air-bubbling and backflushing the membrane and cleaning in place every two months for one or two hours with citric acid, caustic soda or hypochlorite. Although the biomass concentration increased to about 18–20 g $MLSS\,l^{-1}$ a general decline of the membrane performance was not observed. The treatment performance of the membrane bioreactor was better than in conventional sewage works. COD reduction reached 95% at low F/M ratios and was extremely stable. It has to be recalled, however, that the hydraulic retention time of 10.4 h was still comparatively high in this study and will have to be reduced further in future studies.

Nitrification was complete after 30 d. Nitrification should be enhanced by the high temperature in the reactor. Denitrification was only complete after 100 d, but was also relatively stable afterwards. Investigations on whether nitrogen can be eliminated completely by post-denitrification are planned.

Further research will have to focus on phosphorus removal and on kinetics and mass transfer at high sludge concentrations as the biocoenosis and its structure are completely different from the well-known biocoenosis of conventional waste water treatment plants [21]. Increased reaction rates should allow lower hydraulic retention times. Another aspect will be the increasing viscosity of the concentrated sludge, which makes aeration and mixing more difficult.

Complete sludge retention eventually leads to constant sludge concentrations at very low F/M ratios. The mean net sludge production decreased towards zero. Most probably zero net sludge production depends on the composition of the biocoenosis. To be able to control the sludge concentration, investigations of the various influences of the operating conditions on the population in the reactor will be necessary. Although excess sludge production can be suppressed completely, a small amount of sludge will have to be drawn off regularly due to accumulation of inorganic substances in the reactor. However, this amount will be much less than the amount of excess sludge produced in conventional waste water treatment plants.

With the submerged membranes, which are less energy consuming than cross-flow membranes, membrane bioreactors become interesting when a high quality effluent virtually free of particles and microorganisms is needed and when there is only little space.

Acknowledgements

The authors would like to thank Prof. Dr. Dorau of the Federal Environmental Agency for the co-operation and especially for the free use of the membrane bioreactor.

References

[1] EEC Council directive concerning the quality of bathing water (8/12/1975) 76/160/EWG.

[2] Chiemchaisri C, Wong YK, Urase T, Yamamoto K. Organic stabilization and nitrogen removal in a membrane separation bioreactor for domestic wastewater treatment. Water Sci Technol 1992;25:231–40.

[3] Krauth KH, Staab KF. Pressurized bioreactor with membrane filtration for wastewater treatment. Water Res 1993;27:405–11.

[4] Winnen H, Suidan MT, Scarpino PV, Wrenn B, Cicek N, Urbain V, Manem J. Effectiveness of the membrane bioreactor in the biodegradation of high molecular-weight compounds. Water Sci Technol 1996;34:197–203.

[5] Côté P, Buisson H, Pound C, Arakaki G. Immersed membrane activated sludge for the reuse of municipal wastewater. Desalination 1997;113:189–98.

[6] Stephenson T, Judd S, Jefferson B, Brindle K. Membrane bioreactors for wastewater treatment. London: IWA Publishing, 2000.

[7] Yamamoto K, Hiasa M, Mahmood T, Matsuo T. Direct solid-liquid separation using hollow fiber membrane in an activated sludge aeration tank. Water Sci Technol 1989;21:43–54.

[8] Chaize S, Huyard A. Membrane bioreactor on domestic wastewater treatment sludge production, modelling approach. Water Sci Technol 1991;23:1591–600.

[9] Muller EB, Stouthamer AH, van der Verseveld HW, Eikelboom DH. Aerobic domestic waste water treatment in a pilot plant with complete sludge retention by cross-flow filtration. Water Res 1995;29:1179–89.

[10] Low EW, Chase HA. The effect of maintenance energy requirements on biomass production during wastewater treatment. Water Res 1999;33(3):847–53.

[11] Suwa Y, Suzuki T, Toyohara H, Yamagishi T, Urushigawa Y. Single-stage, single-sludge nitrogen removal by an activated sludge process with cross-flow filtration. Water Res 1992;26:1149–57.

[12] Ueda T, Hata K, Kikuoka Y. Treatment of domestic sewage from rural settlements by a membrane bioreactor. Water Sci Technol 1996;34:189–96.

[13] Smith Jr CV, Di Gregorio D, Talcott RM. The use of ultrafiltration membranes for activated sludge separation. Paper Presented at 24th Annual Purdue Industrial Waste Conference, 1969;1300–10.

[14] Bailey AD, Hansford GS, Dold PL. The use of crossflow microfiltration to enhance the performance of an activated sludge reactor. Water Res 1994;28:297–301.

[15] Walter CH. Biological degradation of carbon-, nitrogen-, and phosphorous compounds of a high contaminated waste water in a semi technical three stage reciprocation jet reactor. Dissertation at the Technical University of Berlin, 1989. (in German)

[16] Characklis WG, Marshall K, editors. Biofilms. New York: Wiley; 1990.

[17] Bouillot P, Canales A, Pareilleux A, Huyard A, Goma G. Membrane bioreactors for the evaluation of maintenance phenomena in wastewater treatment. J Ferm Bioeng 1990;69:178–83.

[18] Wenger-Oehn H, Braun R. Examinations of the application of membrane bioreactor systems (German). Abwassertechnik 1994;11/94:49–57.

[19] Pirt SJ. The maintenance energy of bacteria in growing cultures. Proc Roy Soc London 1965;163B:224–31.

[20] Günder B, Krauth KH. Replacement of secondary clarification by membrane separation-results with plate and hollow fibre modules. Water Sci Technol 1999;40: 311–20.

[21] Witzig R, Manz W, Rosenberger S, Krüger U, Kraume M, Szewzyk U. Microbiological aspects of a bioreactor with submerged membranes for aerobic treatment of municipal waste water. Water Res 2001;36:394–402.

PERGAMON

Water Research 36 (2002) 421–428

WATER RESEARCH

www.elsevier.com/locate/watres

Enumeration of acetate-consuming bacteria by microautoradiography under oxygen and nitrate respiring conditions in activated sludge

Jeppe Lund Nielsen, Per Halkjær Nielsen*

Department of Environmental Engineering, Aalborg University, Sohngaardsholmsvej 57, DK-9000 Aalborg, Denmark

Received 29 May 2000; received in revised form 21 February 2001; accepted 22 February 2001

Abstract

Microautoradiography was used to enumerate bacteria able to take up radiolabelled acetate in activated sludge using oxygen or nitrate as electron acceptors. In each of three wastewater treatment plants (WWTP) with nitrification and denitrification (N-removal), the number of bacteria consuming acetate under aerobic and anoxic conditions was identical in contrast to the acetate removal rates. The rates were clearly lower under anoxic conditions suggesting that the specific activity of the cells and not the number of active cells was reduced under anoxic conditions. The fraction of bacteria able to consume acetate varied in three WWTPs between 47% and 93% of the total number of bacteria as determined by DAPI. In a WWTP without N-removal only 20% of the bacteria were able to consume acetate under aerobic conditions and very few of these were able to do it under anoxic conditions. The cell specific acetate removal rates in all WWTPs were found to be $3.0–13.2 \times 10^{-15}\,\mathrm{mol\,cell^{-1}\,h^{-1}}$ under aerobic conditions and between 1.9 and $9.1 \times 10^{-15}\,\mathrm{mol\,cell^{-1}\,h^{-1}}$ under anoxic conditions.

Keywords: Microautoradiography; Active respiring cells; Activated sludge; In situ activity microbial activity; Anoxic reduction factor

1. Introduction

In the activated sludge process for wastewater treatment several microbial processes are of key importance for the treatment performance (e.g. organic removal and nitrogen removal). Many different types of largely unknown and undescribed bacteria carry out these processes. The application of molecular techniques has revealed that the phylogenetical diversity among these bacteria in activated sludge is very large [1]. However, it is still almost impossible to link a certain function to the identity, except for a few distinct, phylogenetic groups, such as ammonia and nitrite oxidisers [2]. Therefore, it is often more convenient to characterise the important bacteria by functional groups carrying out a certain function such as nitrification, denitrification, or uptake of orthophosphate. Information about the population size, activity etc. of a specific functional group can be obtained by measurement of substrate uptake rates, accumulation of metabolites or by the removal of electron acceptors, such as oxygen uptake rate (OUR) or nitrate utilisation rate (NUR) [3].

The measurement of OUR or NUR does not give any information about the number of bacteria involved in the process, which can be valuable knowledge for getting a direct insight into the microbiology of activated sludge. Although several techniques exist which can enumerate and determine the viability of single cells, most methods are not able to detect a certain function by the cells. Only microautoradiography (MAR) can enumerate active bacteria and relate a certain function to individual cells. MAR has been applied mainly in studies of marine systems [4–7]. The method has primarily been used for quantification of the total

*Corresponding author. Tel.: +45-96-35-8080; fax: +45-98-14-25-55.

E-mail address: phn@civil.auc.dk (P.H. Nielsen).

0043-1354/02/$ - see front matter
PII: S0043-1354(01)00224-X

number of active bacteria and its relationship with colony-forming units or total bacterial number estimated by epifluorescent methods such as acridine orange or DAPI staining [8,9]. Recently, MAR has also been applied in activated sludge to study in situ physiology of filamentous bacteria [10], but the method has not yet been used for enumeration studies in this system.

The main disadvantage of using MAR is that there is no single substrate that targets all living microorganisms, thereby detecting all active cells. Mixtures of amino acids or glucose are often used as substrates, but in the work of e.g. [10] it is shown that a substantial fraction of the cells in activated sludge is not able to consume these substrates. However, if the aim is to investigate the number of bacteria consuming a specific substrate, the method is ideal. Another disadvantage of MAR studies is the laboriousness and tediousness of it [8]. However, as we demonstrate in this study, it is possible to simplify the technique and still get reliable and useful results.

It was the objective of this study to modify the MAR procedure and use it for an enumeration study in activated sludge. We present a quantification of two important functional groups of bacteria in activated sludge: bacteria able to consume acetate with oxygen as electron acceptor (under aerobic conditions) and bacteria able to consume acetate with nitrate as electron acceptor (under anoxic conditions). We wanted to determine whether the often observed ratio between OUR and NUR (NUR/OUR) of 0.6–0.8 in electron equivalents [11,12] was due to differences in population sizes or in the specific activity of the cells.

2. Materials and methods

2.1. Activated sludge

Activated sludge samples were obtained from four Danish wastewater treatment plants (WWTP). A description of the four WWTPs is listed in Table 1. All samples were collected from the aeration tanks on the same day as the experiments were conducted.

Table 1
Description of activated sludge treatment plants investigated[a]

WWTP	Process	Sludge age days
Aalborg West	C,N,DN,EBPR,CP	20–30
Aabybro	C,N,DN,CP	20–30
Aalborg East	C,N,DN,EBPR,CP	20–30
Aarestrup	C, (N)	5–10

[a] C: Organic matter removal; N: Nitrification; DN: Denitrification; CP: Chemical phosphorus removal; EBPR: Enhanced biological phosphorus removal; (N): Nitrification in summer periods.

2.2. Incubation conditions for MAR enumeration

The activated sludge was diluted to a final dry matter concentration of approximately 1 g SS/l with effluent water from the same treatment plant. A final volume of 2 ml was transferred to 9 ml glass vials and closed with butyl rubber stoppers. Before addition of the substrate, the vials for anoxic incubations were flushed and evacuated with oxygen-free nitrogen and allowed to stand on a rotary shaker at 150 rpm for 30 min in order to ensure complete removal of traces of oxygen. Nitrate was added to a final concentration of 2 mM. Acetate was added to a final concentration of 1 mM together with 10 μCi of tritium labelled acetate (100 mCi/mmol) (Amersham-Pharmacia Biotech). Strict anaerobic techniques were used for the anoxic incubations. All incubations were conducted at 21°C on a rotary table at 150 rpm for 1 h. Some samples were incubated for 2 and 3 h in order to test that the incubation time was sufficient. Aerobic incubations were incubated horizontally to prevent diffusion-limited conditions. After the incubation period the samples were fixed by adding paraformaldehyde (final concentration 4%) at 4°C for 2–3 h. The fixed sludge was washed 4 times (each 8 min centrifugation at 3400 × g) in tap water to ensure removal of excess radioactive substrate. After washing the suspension was homogenised using a glass tissue grinder (Thomas Scientific®, USA). Dilutions were made in sterile, filtered tap water (0.22 μm polycarbonate filter, Millipore). Controls with pasteurised sludge samples (10 min at 80°C) or samples without addition of tracer were included in order to examine for excitation of the silver ions in the photographic film by non-radioactive chemical compounds (chemography) and other artefacts. Such artefacts were never observed.

2.3. Microautoradiographic procedure

The fixed, homogenised activated sludge samples were diluted in sterile, filtered water and 1 ml subsamples of an appropriate dilution were filtered onto a 0.22 μm white polycarbonate membrane filter (Millipore). The filters were placed on microscope slides and covered by a liquid film emulsion (LM-1 from Amersham-Pharmacia Biotech) by gentle dipping in pre-warmed emulsion (43°C). The slides were exposed in the dark at 4°C for approximately four weeks. Samples incubated under aerobic conditions from each of the WWTPs tested were used to test for the optimal exposure time, and a slide was developed every week to follow the development in number of cells covered with silver grains. Development was carried out using Kodak D19 developer [10].

2.4. Enumeration by microscopy

The MAR-positive cells on the polycarbonate filters were counted using bright field microscopy. Total number of cells was determined by epifluorescence microscopy after staining by DAPI (4,6-diamino-2-phenylindoldihydrochlorid-dilactate) on separate filters (black 0.22 μm polycarbonate filters, Millipore). In some experiments a combination of bright field microscopy and epifluorescence microscopy was used for simultaneous detection of MAR-positive bacteria and total number of bacteria stained by DAPI on samples placed on cover glass [13]. DAPI staining was applied after fixation of the sludge with paraformaldehyde by adding DAPI to a final concentration of 1 mg/ml for 30 min in the last washing step. Visualisation of the combined fluorescent signals and bright field microscopy were performed using an inverted laser-scanning microscope (Zeiss, LSM 510) as described by [13]. A MAR-positive cell was defined as an assembly of more than 5 silver grains covering an area of 2–3 μm^2, while the background grain density was less than 0.01 silver grain per μm^2. The number of MAR-positive cells and total DAPI counts was estimated by counting a minimum of 2500 silver grain covered bacteria or DAPI stained bacteria on each filter, with the exception of the numbers obtained from Aarestrup WWTP, where 450 silver grain covered cells were counted. Average cell numbers are presented for duplicate or triplicate experiments. Standard error on the mean was less than 8%. Lysis of the bacterial cells during the homogenisation procedure did not take place as no activity of the intracellular enzyme, glucose-6-phosphate dehydrogenase, was found in the cell-free supernatant after homogenization [14].

2.5. Acetate consumption rates

A parallel flask with activated sludge was used for the measurement of acetate consumption rates and as a control for proper incubation conditions. Intermittent sampling during the incubation was carried out by chilling a subsample to 0–1 °C, centrifuging at 10,000 × g for 10 min, and filtrating the supernatant through a 0.2-μm polycarbonate filter before freezing the sample for later analysis by HPLC. Furthermore, in these studies a possible depletion of the substrate, electron acceptor, and/or accumulation of nitrite under the anoxic experiment were recorded. Uptake of the labelled acetate into the biomass was followed by scintillation counter (Packard model 1600 TR). A pellet obtained by centrifugation of 1 ml sample (8 min at 6000 × g) was resuspended in tap water, and 100–200 μl of this sample was added to 2 ml scintillation liquid (Ultima Gold XR; Packard Instr. Co.) in the dark for at least 10 h before counting.

2.6. Analytical techniques

Acetate removal was measured by a Dionex ion chromatograph equipped with IonPac AS11 high capacity column and a suppressed conductivity detector using 0.2 mM NaOH as eluent. Formation of nitrite was analysed by autoanalyser (Technicon TRAACS 800 autoanalyser). Suspended solids (SS) and organic content (volatile suspended solids, VSS) were measured according to [15].

3. Results

3.1. The MAR procedure

All enumerations of acetate-consuming bacteria by MAR were carried out after filtration of diluted, homogenised sludge samples onto polycarbonate filters. The total number of bacteria was determined on separate filters after DAPI staining. In order to ensure that all MAR-positive cells were indeed bacteria, DAPI staining was combined with MAR on cover glasses and microscopically examined through the cover glass. All cells covered with silver grains were also stained by DAPI confirming that no artefacts were present. In some samples of activated sludge (from Aalborg West WWTP) the two methods were compared by estimating the ratio of MAR-positive cells to the total DAPI count on cover glass and comparing it to MAR and DAPI determinations on separate polycarbonate filters. Almost identical results were obtained by these two approaches (Table 2).

Fig. 1 shows how the silver grains were located on top of the bacteria on a polycarbonate filter, resulting in single cell resolution. Assemblies of silver grains were located less than 1 μm from the cell surface. Despite careful homogenisation of the samples, a small fraction of the cells were assembled into cell-aggregates, but it was still possible to distinguish the single cells by MAR as most of the cell-aggregates were constituted by cells that were not covered by silver grains. These aggregates most probably were autotrophic nitrifying cells, as these were present as microcolonies in the activated sludge as assessed by fluorescence in situ hybridisation with oligonucleotide probes (FISH) (unpublished results).

The radiolabelled activated sludge samples were exposed until all cells with uptake of acetate were visible. This was determined by counting the number of MAR-positive cells appearing during the exposure period by developing a sample every week. After 3–4 weeks the number of MAR-positive cells did not increase further, and at this time all samples were developed and counted. The number of silver grains on top of the cells did not increase any further with longer exposure time. Furthermore, a prolonged incubation

Table 2
Enumeration of acetate-consuming cells in activated sludge with oxygen or nitrate as electron acceptor by using DAPI count for total enumeration and microautoradiography for enumeration of acetate-consuming cells

WWTP	Electron acceptor	Total DAPI count (cells ml^{-1})	MAR (cells ml^{-1})	MAR/Total (%)	Ratio (NO_3^-/O_2) (%)
Aalborg West	O_2	0.66×10^9	0.48×10^9	73.6	96.2
	NO_3^-	1.72×10^9	1.22×10^9	70.9	
Aalborg West	O_2	N/A	N/A	75.0[a]	97.7
	NO_3^-	N/A	N/A	73.3[a]	
Aabybro	O_2	1.43×10^9	1.28×10^9	89.2	104.3
	NO_3^-	2.08×10^9	1.94×10^9	93.1	
Aalborg East	O_2	2.77×10^9	1.29×10^9	46.7	113.2
	NO_3^-	2.13×10^9	1.13×10^9	52.9	
Aarestrup	O_2	0.58×10^9	0.11×10^9	19.2	5.2
	NO_3^-	0.76×10^9	7.66×10^6	1.0	

[a] Obtained from simultaneous count of MAR-positive and DAPI stained cells on cover glass.
N/A: not applicable.

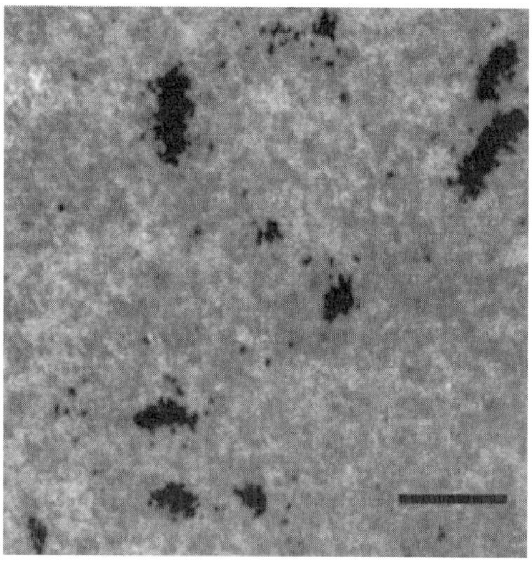

Fig. 1. Bright field image of single cells positive in taking up [^{3}H]acetate on a polycarbonate filter, showing the resolution of the silver grains on top of the bacteria. Bar equals 5 μm.

time up to 3 h did not increase the number of MAR-positive cells, indicating that active cells had taken up enough detectable labelled acetate during the applied incubation period of 1 h.

The removal of [^{3}H]acetate from the bulk phase and incorporation into the biomass is shown in Fig. 2. The acetate uptake rate was constant throughout the incubation period, and no initial lag phase was observed under either of the electron acceptor conditions investigated. The average amount of labelled acetate incorporated into each MAR-positive cell during the incubation period was between $0.38–1.1\times10^{-15}$ Ci ($1.4–4.1\times10^{-5}$ Bq). This corresponded to an average of 28–76 disintegration per MAR-positive cell in the exposure time applied of 3–4 weeks.

3.2. Enumeration of acetate-consuming bacteria

The number of aerobic acetate-consuming bacteria ranged from 19–89% of the total number of bacteria in the 4 WWTPs (Table 2), suggesting variability in the presence of bacteria able to consume acetate. WWTPs with N-removal (nitrification and denitrification) had relatively high numbers of acetate-utilising bacteria (47–93%). An important observation was that the number of acetate-consuming bacteria under both aerobic and anoxic conditions was almost identical in each treatment plant, indicating presence of mainly facultative, denitrifying bacteria (Table 2).

In the WWTP without N-removal, Aarestrup WWTP, only 19% of the total count could consume acetate with oxygen as the electron acceptor while hardly any bacteria were able to consume acetate under anoxic conditions (only 1%, Table 2). This corresponded well to the absence of a nitrification–denitrification step in this WWTP.

3.3. Acetate consumption rates

The acetate removal rate was higher under aerobic conditions than under anoxic conditions in all WWTPs investigated (Table 3). In WWTP with N-removal, acetate was removed under aerobic conditions with rates from 0.49 to 0.67 mmol g VSS^{-1} h^{-1}. The corresponding acetate removal rates under nitrate-reducing conditions were 52–86% of the aerobic rate. Fig. 2 shows an example of how the acetate was consumed and incorporated into the biomass. The removal of acetate proceeded at a constant rate also after a pre-incubation step of unlabelled acetate for a period of almost 2 h.

In the WWTP without N-removal (Aarestrup) the acetate removal rate was low under aerobic conditions, only 0.3 mmol g VSS^{-1} h^{-1}. In accordance with the plant

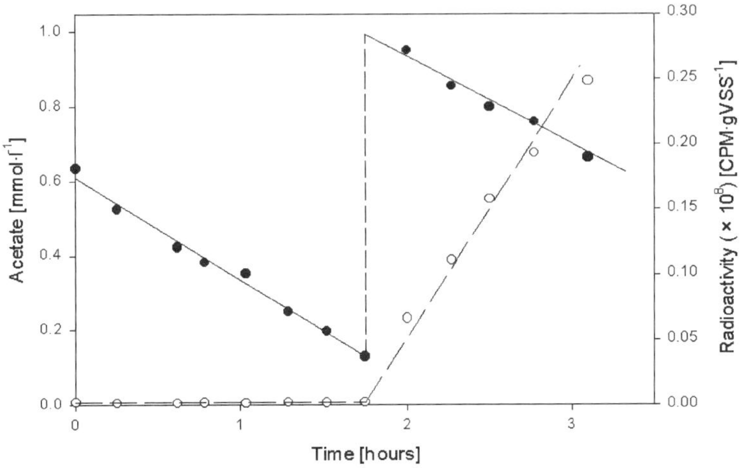

Fig. 2. Removal of acetate (●) in activated sludge and the incorporation of [^{3}H]acetate (○) into the biomass. Dotted line at 105 min indicates start of incubation with addition of acetate and [^{3}H]acetate.

Table 3
Comparison of acetate removal rates in activated sludge from different wastewater treatment plants with nitrate and oxygen as electron acceptors[a]

WWTP	Electron acceptor	Acetate removal rates (mmol/g VSS h)	Ratio (NO_3^-/O_2) (%)
Aalborg West	O_2	0.594	86
	NO_3^-	0.514	
Aabybro	O_2	0.673	52
	NO_3^-	0.350	
Aalborg East	O_2	0.489	77
	NO_3^-	0.377	
Aarestrup	O_2	0.301	21
	NO_3^-	0.062	

[a] Organic matter content was determined to be (in g VSS/l): Aalborg West (3.27), Aabybro (3.26), Aalborg East (3.68) and Aarestrup (1.06).

operation, the acetate removal rate under denitrifying conditions was very low (Table 3).

Nitrite was produced and accumulated during the anoxic incubations from all four WWTPs at a constant rate resulting in a maximum of 0.16 mM after 1 h of incubation. No accumulation of nitrite was measured in aerobic incubations. Nitrate was not depleted during the anoxic incubations.

Average cell-specific acetate uptake rates were calculated from the numbers of acetate-consuming bacteria and the acetate removal rates. These rates ranged from $3.0–13.2 \times 10^{-15}$ mol cell^{-1} h^{-1} for aerobic uptake of acetate and slightly lower, between $1.9–9.1 \times 10^{-15}$ mol cell^{-1} h^{-1} for anoxic uptake (Table 4).

Table 4
Comparison of cell-specific acetate uptake rates determined with nitrate or oxygen as electron acceptor

WWTP	Electron acceptor	MAR (cells g VSS^{-1})	Specific rate (mol cell^{-1} h^{-1})
Aalborg West	O_2	1.48×10^{11}	13.2×10^{-15}
	NO_3^-	1.86×10^{11}	9.01×10^{-15}
Aabybro	O_2	3.93×10^{11}	5.59×10^{-15}
	NO_3^-	5.94×10^{11}	1.92×10^{-15}
Aalborg East	O_2	3.51×10^{11}	5.12×10^{-15}
	NO_3^-	3.07×10^{11}	4.53×10^{-15}
Aarestrup	O_2	1.05×10^{11}	3.04×10^{-15}
	NO_3^-	0.07×10^{11}	9.13×10^{-15}

Thus, in the different WWTPs, the cell-specific rate under anoxic conditions was only between 34% and 88% of the specific rate under aerobic conditions. It was also observed that, although the number of acetate consumers at Aarestrup WWTP was very small, the cell-specific rates were similar to those of the other WWTPs.

4. Discussion

4.1. The MAR procedure

In this study we have shown that MAR can be a very useful method for quantification of bacteria capable of utilising a specific organic substrate under distinct electron acceptor conditions in activated sludge. This technique can be used to quantify other functional groups of importance to wastewater treatment as well as studies of in situ physiology [10,16]. Application of

MAR for enumeration of functional groups in activated sludge can either be performed on cover glass or conducted after filtration onto a polycarbonate filter. The cover glass approach is valuable as a control for a direct localisation of the MAR-positive bacteria or, if desirable, to measure the ratio of active bacteria compared to the total number of bacteria in the sample. This modification of the original method for combining MAR and DAPI [8] makes the method relatively easier to use. Given the control on cover glass and controls for chemography, MAR can be used on filters for a direct enumeration of active bacteria without any special problems.

A sufficient homogenisation of microcolonies in the flocs can under certain circumstances be a problem for a proper enumeration. We found that almost all bacteria present in clusters after homogenisation were unable to take up acetate under aerobic and anoxic conditions, so they were probably autotrophic cells. Autotrophic bacteria in activated sludge are usually found in dense clusters [2], and it is our experience that these cell clusters hardly disrupt upon the applied homogenisation, so enumeration of these bacteria (by fixation of labelled bicarbonate) can be problematic. For the heterotrophic bacteria enumerated in this study, the similar numbers obtained by the two approaches (filter and coverglass) indicates that a good estimate of the total number of acetate consuming heterotrophic cells were obtained.

To be considered MAR-positive, a cell must take up sufficiently radiolabelled substrate during the incubation period to produce enough silver grains on the autoradiograph under given exposure conditions. The amount of tracer taken up in each cell is dependent on the amount of labelled and unlabelled substrate added, substrate uptake rate, and the incubation time. In the experiments presented here, approximately 10^{-15} Ci were incorporated into each cell. The subsequent exposure period required was around 4 weeks. This period could be reduced, e.g. to 3–6 days by adjusting any of the above-mentioned parameters so more tracer would be incorporated into the cells.

Application of a pre-incubation step with unlabelled acetate for almost 2 h showed no effect on the acetate consumption rate as shown in Fig. 2. It did also not change the number of MAR-positive bacteria (results not shown) showing that acetate was not just transiently taken up and accumulated as storage compounds. Thus, only bacteria with a continuous uptake on acetate were scored as MAR-positive.

The number of silver grains covering the bacteria was very similar for most cells, which roughly suggests that a similar amount of tracer was taken up. A small percentage of the cells did, however, only become MAR-positive after 3–4 weeks of exposure, indicating a low level of labelling and thus a low substrate uptake activity. However, as MAR is not considered a quantitative method in this type of studies [16], variations in silver grain density were not further investigated.

4.2. Acetate-consuming bacteria in activated sludge

The difference between the aerobic and anoxic removal rate of COD in activated sludge has previously been reported to reflect the fact that only a certain fraction of the aerobic, heterotrophic population was also active under anoxic conditions [11]. In models of the activated sludge process, this anoxic reduction factor is, according to [12], typically 0.6, but must be determined for each system investigated. The results presented here clearly show that the anoxic reduction factor in the investigated WWTPs with N-removal was due to a general reduction in cell activity and not due to a reduction in number of active cells. This supports the findings by [17–19] who in both pure and enriched cultures found the energy production efficiency with oxygen as electron acceptor to be reduced with approximately 40% when nitrate is electron acceptor.

The identical number of aerobic and anoxic acetate-consuming bacteria in each of the N-removing WWTPs investigated was not due to two different populations of the same size as up to 93% of the total number of bacteria were able to take up acetate under both conditions. This suggests a vast overlap of facultative, denitrifying bacteria able to perform both aerobic and anoxic acetate uptake.

The average cell-specific acetate removal rate under anoxic conditions was reduced to 34–88% of the aerobic rate for the three WWTPs with N-removal. This is in agreement with the fact that oxygen is energetically more favourable as electron acceptor than nitrate resulting in a decrease in the cell-specific rate. The average cell-specific uptake rate of acetate was found to be very similar among the four WWTPs investigated. These rates are comparable to values that can be calculated from pure culture studies of [20,21] to approximately 10^{-15} mol cell^{-1} h^{-1}, assuming a mean cell size of 0.3 μm^3 and a dry matter content of 25%.

The fraction of bacteria able to consume acetate varied in the different treatment plants. In two plants nearly all cells were able to consume acetate, indicating that nearly all cells were living and active. The fact that they all consumed acetate is in agreement with the general high content of acetate in most wastewaters [22]. The fraction of acetate consumers was significant lower in the other plants investigated. However, the DAPI staining also includes non-viable cells, anaerobic bacteria, and autotrophic bacteria, so the lower fraction of acetate consumers cannot be explained unless additional investigations with uptake of other substrates and a quantification of autotrophs are conducted.

The nitrifying bacteria, which are often found in dense cell clusters, might be slightly underestimated during the enumeration. However, as only around 5% of all Eubacteria was found to be nitrifiers, based on in situ hybridisation, these autotrophic cells will not significantly overestimate the fraction of heterotrophic acetate consuming bacteria. Furthermore as very similar cell specific rates were found in the different WWTPs under the conditions tested here, clustering of certain cells seems not to influence the ratio of MAR positive cells to the total DAPI count.

The treatment plant without N-removal had a surprisingly low number of aerobic acetate-consuming bacteria and an almost complete absence of denitrifiers. However, the MAR-enumeration seems correct because the cell specific rates were the same as for the other plants. The very low number of acetate consumers can only be explained by a more detailed investigation of the treatment plant and the wastewater composition.

Today MAR is the only available method to quantify the bacterial number carrying out a certain function in a complex microbial system such as activated sludge. Recently, it has also been described how MAR can be combined simultaneously with an identification of the MAR-positive cells by fluorescently labelled oligonucleotide probes [13,16]. This combination has a great potential to detect and investigate the diversity and community composition of various functional groups in activated sludge in future studies.

5. Conclusion

The MAR was used successfully to enumerate acetate-consuming bacteria in activated sludge under aerobic and anoxic conditions. In each of 3 WWTP with N-removal investigated, the number of bacteria able to consume acetate was identical under aerobic and anoxic conditions. The high fraction of acetate-consuming bacteria (up to 93% of the total number) indicated that the vast majority were facultative nitrate-reducing bacteria.

The removal rate of acetate under anoxic conditions was only 30–80% of the rate under aerobic conditions. Because the population of acetate-consuming nitrate reducers was of the same size as the aerobic population, the lower acetate removal rate under anoxic conditions was due to a lower specific cell activity and not to a reduced number of active cells.

In a treatment plant without N-removal, the number of aerobic acetate-consuming bacteria as well as the acetate removal rates was low. Furthermore, very little acetate was taken up under anoxic conditions where also the number of active bacteria detected by MAR was very low.

For all treatment plants investigated, the average cell-specific rate for acetate uptake was found to be in the range: $3.0–13.2 \times 10^{-15}$ mol cell^{-1} h^{-1} for aerobic uptake of acetate, and between $1.9–9.1 \times 10^{-15}$ mol cell^{-1} h^{-1} for anoxic uptake of acetate.

Acknowledgements

M. Stevenson and M. Fredsgaard are thanked for valuable technical assistance. The Danish Technical Research Council funded the study by the framework programme "Activity and Diversity in Complex Microbial Systems".

References

[1] Bond PL, Hugenholtz P, Keller J, Blackall LL. Bacterial community structures of phosphate-removing and non-phosphate-removing activated sludges from sequencing batch reactors. Appl Environ Microbiol 1995; 61(5):1910–6.

[2] Juretschko S, Timmermann G, Schmid M, Schleifer K-H, Pommerening-Röser A, Koops H-P, Wagner M. Combined molecular and conventional analyses of nitrifying bacterium diversity in activated sludge: *Nitrosococcus mobilis* and *Nitrospira*-like bacteria as dominant populations. Appl Environ Microbiol 1998;64(8):3042–51.

[3] Kristensen GH, Jørgensen PE, Henze M. Characterization of functional microorganism groups and substrate in activated sludge and wastewater by AUR, NUR and OUR. Water Sci Tech 1992;25(6):43–57.

[4] Brock ML, Brock TD. The application of micro-autoradiographic techniques to ecological studies. Mitt Int Ver Limnol 1968;15:1–29.

[5] Hoppe H-G. Determination and properties of actively metabolizing heterotrophic bacteria in the sea, investigated by means of micro-autoradiography. Mar Biol 1976;36:291–302.

[6] Tabor PS, Neihof RA. Direct determination of activities for microorganisms of Chesapeake Bay populations. Appl Environ Microbiol 1984;48(5):1012–9.

[7] Pedrós-Alió C, Newell SY. Microautoradiographic study of thymidine uptake in brakish waters around Sapelo Island, Georgia, USA. Mar Ecol Prog Ser 1989; 55:83–94.

[8] Meyer-Reil L-A. Autoradiography and epifluorescence microscopy combined for the determination of number and spectrum of actively metabolizing bacteria in natural waters. Appl Environ Microbiol 1978;36(3): 506–12.

[9] Grossmann S. Bacterial activity in sea ice and open water of the Weddel Sea, Antarctica: a microautoradiographic study. Micro Ecol 1994;28(1):1–18.

[10] Andreasen K, Nielsen PH. Application of microautoradiography to study substrate uptake by filamentous microorganisms in activated sludge. Appl Environ Microbiol 1997;63(9):3662–8.

[11] Henze M, Grady Jr. CPL, Marais GvR, Matsuo T. Activated Sludge Model No. 1. IAWPRC. Scientific and Technical Report No. 1, IAWPRC, London, 1987.

[12] Gujer W, Henze M, Mino T, van Loosdrecht MCM. Activated sludge Model no. 3. Water Sci Tech 1999; 39(1):183–93.

[13] Lee N, Andreasen K, Nielsen PH, Juretschko S, Nielsen JL, Schleifer K-H, Wagner M. Combination of fluorescent in situ hybridization and microautoradiography—a new tool for structure-function analysis in microbial ecology. Appl Environ Microbiol 1999;65(3):1289–97.

[14] Frølund B, Palmgren R, Keiding K, Nielsen PH. Extraction of extracellular polymers from activated sludge using a cation exchange resin. Water Res 1996;30(8):1749–58.

[15] Standard Methods. Standard Methods for the examination of water and wastewater. 19th ed. New York, NY, USA: American Public Health Association, American Water Works Association and Water Pollution Control Federation, 1995.

[16] Nielsen PH, Andreasen K, Lee N, Wagner M. Use of microautoradiography and fluorescent in situ hybridization for characterization of microbial activity in activated sludge. Water Sci Tech 1999;39(1):1–9.

[17] John P, Whatley FR. Oxidative phosphorylation coupled to oxygenuptake and nitrate reduction in *Micrococcus denitrificans*. Biochim Biophys Acta 1970;216:342–52.

[18] van Verseveld HW, Bosma G. The respiratory chain and energy conservation in the mitochondrion-like bacterium *Paracoccus denitrificans*. Microbiol Sci 1987;4(11):329–33.

[19] Kuba T, Murnleitner E, van Loosdrecht MCM, Heijnen JJ. A metabolic model for biological phosphorus removal by denitrifying organisms. Biotechnol Bioeng 1996;52(6):685–95.

[20] Tempest DW, Neijssel OM. The status of Y_{ATP} and maintenance energy as biological interpretable phenomena. Ann Rev Microbiol 1984;38:459–86.

[21] Tempest DW, Neijssel OM. Physiological and energetics aspects of bacterial metabolite overproduction. FEMS Microbiol Lett 1992;100:169–76.

[22] Raunkjær K, Hvitved-Jakobsen T, Nielsen PH. Transformation of organic matter in a gravity sewer. Water Environ Res 1995;67(2):181–8.

PERGAMON

Water Research 36 (2002) 429–438

WATER RESEARCH

www.elsevier.com/locate/watres

Microbial community analysis of thermophilic contact oxidation process by using ribosomal RNA approaches and the quinone profile method

Futoshi Kurisu[a,*], Hiroyasu Satoh[b], Takashi Mino[b], Tomonori Matsuo[c]

[a] *Research Center for Water Environment Technology, Graduate School of Engineering, the University of Tokyo, 7-3-1 Hongo, Bunkyo-ku, Tokyo 113-8656, Japan*

[b] *Institute of Environmental Studies, Graduate School of Frontier Science, the University of Tokyo, 7-3-1 Hongo, Bunkyo-ku, Tokyo 113-0033, Japan*

[c] *Department of Civil and Environmental Engineering, Toyo University, 2100 Kujirai, Kawagoe-City, Saitama 350-8585, Japan*

Received 14 June 2000; received in revised form 2 January 2001; accepted 23 January 2001

Abstract

Microbial community structure of a lab scale thermophilic aerobic wastewater treatment reactor was analyzed by a combination of culture-independent methods. Quinone profile method provides for chemical analysis of respiratory quinone molecular species, which corresponds to bacterial groups. Denaturing gradient gel electrophoresis (DGGE) of PCR-amplified 16S rDNA partial sequences (PCR-DGGE) clarifies community changes at species level, as DGGE can separate DNA fragments of different sequences. Certain phylogenetic groups of bacterial cells can be labeled by fluorescence in situ hybridization (FISH). Quinone profile showed a predominant presence of MK-7. PCR-DGGE revealed that constituents of the community were unchanged during the stable phase. FISH demonstrated the existence of the relatives of *Bacillus lentus* and *B. thermocloacae* in considerable proportions. The community was mainly composed of *Bacillaceae*, and obligate thermophilic and mesophilic *Bacillus* appeared in spite of the temperature fluctuation from 35°C to 60°C. The combination of these culture-independent methods revealed the community precisely enough to evaluate the reactor performance.

Keywords: Microbial community structure; Quinone profile; PCR-DGGE; Fluorescence in situ hybridization; Thermophilic contact oxidation process; Thermophilic *Bacillus*

1. Introduction

Thermophilic contact oxidation process (TCOP) is a novel method for the treatment of high strength organic wastewater. In TCOP, wastewater is aerobically mixed with wood chips. The wood chips absorb wastewater and the organic compounds in the wastewater are aerobically degraded on or in the wood chips. In this process, most of the organic carbon is oxidized down to carbon dioxide and the resulting degradation heat increases the reactor temperature up to 50–70°C. Water is evaporated due to aeration under high temperature. However, occasionally the temperature does not increase to thermophilic range from the startup, or after a short period of operation. Therefore, it is necessary to elucidate the mechanism of the treatment from the perspective of microbial community in relation to the physico-chemical conditions.

Conventional culture-dependent methods are inadequate for community analysis due to their heavy bias through cultivation [1]. Recently developed techniques based on chemotaxonomy and molecular biology overcome this problem. However, no single method is sufficient to analyze complex communities present in

*Corresponding author. Tel.: +81-3-5841-6238; fax: +81-3-5841-8529.

E-mail address: kurisu@env.t.u-tokyo.ac.jp (F. Kurisu).

0043-1354/02/$ - see front matter
PII: S0043-1354(01)00225-1

the biological wastewater treatment processes as quantitative and detailed information of the total community is necessary to characterize the process performance.

A common chemotaxonomical method is the quinone profile procedure, which entails direct analysis of respiratory quinones in cell membranes to quantitatively reveal community profiles according to quinone molecular types [2]. Quinone molecular types can be used to describe phylogenetic groups. However, quinone profiles can only reveal the structure at 'phylum' level and some phylogenetically different groups of bacteria share the same quinone groups [3]. Fluorescence in situ hybridization (FISH) with oligonucleotide probes directed against rRNA can more specifically quantify the proportion of microbial groups of interest, from domain level to species level [4]. However, groups that can be quantified by FISH are limited to those to which probes are available.

Denaturing gradient gel electrophoresis (DGGE) of PCR-amplified gene fragments coding for 16S rRNA (PCR-DGGE method) has recently been utilized to determine the genetic diversity of natural microbial communities [5]. By DGGE, DNA fragments of the same length but with different base pair sequences, such as PCR fragments obtained from a mixed culture, can be separated. Similarity of microbial communities can be analyzed from electrophoresis patterns and bands can further be excised from gels and sequenced to identify the phylogenetic affiliation of the community members [6].

In the present study, a TCOP microbial community was analyzed by a combination of culture-independent methods. A bench scale TCOP reactor was operated using peptone and starch as primary carbon sources. An overview of the community at the group level was provided by the respiratory quinone profile method and by FISH. PCR-DGGE was utilized to determine population dynamics at species level. DGGE bands were excised and sequenced to elucidate species that constituted the community. Finally, FISH was carried out with oligonucleotide probes specific to the DGGE bands to quantify their proportion. According to the results of community analysis, community structure and its variation were assessed in relation to process performance.

2. Materials and methods

2.1. Operation and monitoring of TCOP reactor

The experimental setup is shown in Fig. 1. The substratum consisted of 6 l of wood chips sized 2–4 mm. To maintain a constant rate of heat loss, the reactor was placed in a polystyrene foam box and the temperature outside the reactor was controlled to be 4°C lower than inside. The wood chips inside the reactor were mixed at the rate of 1 rpm with the aid of a mixing blade.

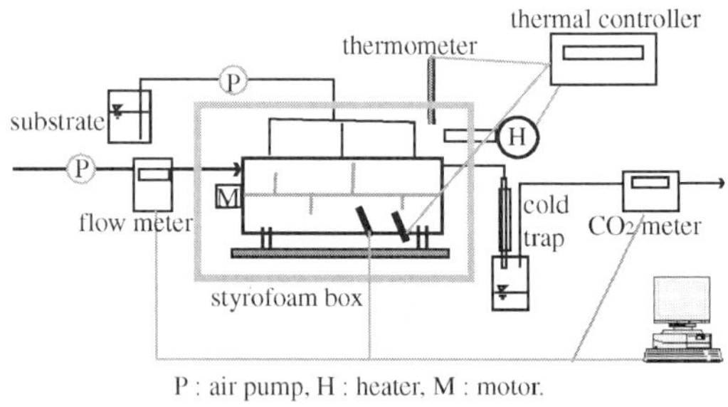

Fig. 1. The experimental setup of the reactor.

The reactor was fed with 300 ml of synthetic wastewater every 48 h. The synthetic wastewater contained $82\,g\,l^{-1}$ of soluble starch, $82\,g\,l^{-1}$ of peptone, $41\,g\,l^{-1}$ of sucrose and $41\,g\,l^{-1}$ of yeast extract. The carbon and nitrogen contents were $11\,g\,C\,l^{-1}$ and $1.6\,g\,N\,l^{-1}$, respectively. At the beginning of the operation, the reactor was inoculated with 60 g (dry weight) of semi-aerobic compost. The reactor was aerated via a mass flowmeter at a rate of $1.8–2.0\,l\,min^{-1}$ ($300–370\,l\,[m^3\ chip]^{-1}\,min^{-1}$). The carbon dioxide concentration in emission gas was monitored by CO_2 meter (LX-710, Iijima electronics, Aichi, Japan). Data of CO_2 concentration, aeration rate and temperature were acquired by means of a personal computer once every minute.

2.2. Sampling

Samples for community analysis were taken from the reactor at the times shown in Fig. 2. For quinone analysis, approximately 5 g of wood chips were taken from the reactor and stored at −20°C for up to a month. For PCR-DGGE and FISH, 0.5 g of the samples were immersed in 5 ml sterilized 1 × phosphate-buffered saline (PBS; pH 7.2) and the immersion was sonicated for 5 min at 15 W (Branson Sonifier 450, Krautkramer-Branson, Lewistown, Penn.). The suspension was filtered with a coarse filter paper (No. 1, Advantec-Toyo, Tokyo, Japan) to remove large particles and was stored in liquid nitrogen until use.

2.3. Quinone profile

Frozen samples were dehydrated by lyophilization before quinone extraction. Lipid fraction including quinone was extracted with a chloroform–methanol (2 : 1, vol/vol) mixture. The crude extract was purified and separated to menaquinone (MK) and ubiquinone (Q) by solid phase extraction using Waters Sep-Pak silica (Millipore-Japan, Tokyo, Japan) as described by Fujie et al. [7]. Purified quinones were then analyzed by reverse phase HPLC (HP 1100, Hewlett-Packard Japan, Tokyo, Japan) equipped with a Waters Novapac C18 (Millipore-Japan) column and a photodiode-array detector. For HPLC analysis, a mixture of methanol and 2-propanol (9 : 2, vol/vol) was used as mobile phase, the column was kept at 45°C and then UV absorbance was measured (MK: 270 nm, Q: 275 nm). Molecular types of quinone were identified by comparing elution times with those of standard quinones as well as by measuring molecular weight by mass spectrometry (LCQ, Finnigan, ThermoQuest, Tokyo, Japan) following HPLC. The percentage of each quinone component was calculated using the peak area of each component and the mole absorbance of MK and Q ($17.4\,mM^{-1}\,cm^{-1}$ [at 270 nm] and $14.4\,mM^{-1}\,cm^{-1}$ [at 275 nm], respectively).

2.4. DNA extraction and PCR amplification

One milliliter of suspended samples was washed with TE buffer (pH 8) and genomic DNA was extracted with benzyl chloride as described by Zhu et al. [8]. Primers were selected to amplify a part of 16S rDNA, positioning 341 to 534 in *E. coli* position. The nucleotide sequence of the forward primer, which is specific to eubacteria, contains at its 5′ end a 40-base GC clamp (5′-CGCCCGCCGCGCGCGGCGGGCGGGGCGGGGGCACGGGGGG-3′) (357fGC; 5′-(GC-clamp)-CCTACGGGAGGCAGCAG-3′) [5]. The universal consensus sequence was used as a reverse primer (518r; 5′-ATTACCGCGGCTGCTGG-3′). PCRs were performed using GeneAmp 9600 thermalcycler (Applied Biosystems, Perkin Elmer Japan, Tokyo, Japan) and AmpliTaq Gold DNA polymerase (Applied Biosytems). PCR mixtures were preincubated at 94°C for 9 min. Denaturing, annealing and extension were carried out at 94°C, 53°C and 72°C, respectively and the duration was 30 s in all the steps. This cycle was repeated 35 times, and then incubated at 72°C for 10 min for the final elongation.

2.5. Analysis of PCR products by DGGE

DGGE was performed according to Muyzer et al. [5] with small modifications, as described elsewhere [9]. The 8% polyacrylamide gel was used for DGGE, and the denaturant gradient of urea and formamide was 35–50%. The gels were run by D Code system (Bio-Rad Laboratories, Hercules, Calif.) for 300 min at 60°C and 130 V.

After completion of electrophoresis, the gels were stained with Vistra Green (Amersham Pharmacia Biotech, Tokyo, Japan) and documented by fluorescent image scanner (Fluorimager 595, Molecular Dynamics, Sunnyvale, Calif.). Intensely stained bands were excised from the gels and were immersed in 50 μl of sterilized water. DNA was recovered from the gels by freeze-thawing more than 3 times.

2.6. Sequencing

DNA fragments recovered from the DGGE bands were reamplified with the forward primer 357f including an additional sequence extension (T3; 5′-AAAATTAACCCTCACTAAAG-3′) at its 5′ end and the reverse primer 518r with an extension (M13r; 5′-AAATTCACACAGGAAACAG-3′) at its 5′ end to facilitate DNA sequencing [6]. PCR conditions were identical to the first amplification except for the cycle number, which was reduced to 25 times in order to reduce the amplification of contaminants. The newly obtained PCR products were directly sequenced with an SQ-5500 automatic sequencing system (Hitachi, Tokyo, Japan) with T3 and M13r primers. Homology retrieval of the sequences obtained from the DGGE bands was performed by BLAST search option in the DNA Database in Japan (DDBJ) nucleotide sequence database.

2.7. FISH

Suspended samples were fixed with 3% paraformaldehyde for 3 h and stored in a 1 : 1 mixture of 1 × PBS and 99% ethanol at −20°C, as described by Amann [10]. We also tested 50% ethanol for fixation [11] but no substantial difference was observed in the fluorescent signal. For example, FISH to monocultures closest to the DGGE bands from TCOP samples, such as *B. lentus*, *B. thermocloacae*, *Staphylococcus sciuri*, etc., showed no difference in fluorescent intensity (data not shown). The samples were immobilized on slideglasses by air-drying followed by ethanol dehydration.

The oligonucleotide probes used in this study are listed in Table 1. They were synthesized and labeled with X-rhodamine isothiocyanate (XRITC) or Cy5 by TaKaRa Biomedicals (Tokyo, Japan). In situ hybridizations were performed as described by Amann [10]. To quantify the proportion of target species to the total bacteria, probe EUB338 was also hybridized with the samples. Formamide concentration in the hybridization buffer is shown in Table 1. Following the hybridization, slides were washed for 20 min at 48°C in washing buffer without formamide. The NaCl concentrations of the buffer were 730, 450, 170 and 40 mM for probes hybridized with 5%, 10%, 20% and 35% formamide, respectively, which gave equivalent stringency according to Lathe's equation [12]. Subsequently, the washed and dried slides were mounted with Slowfade-light antifade reagent (Molecular Probes, Leiden, the Netherlands). The slides were examined by confocal laser scanning

Table 1
Oligonucleotide probes used in this study

Probe	OPD name[a]	Sequences (5′–3′)	Formamide (%)	References
EUB338	S-D-Bact-0338-a-A-18	GCTGCCTCCCGTAGGAGT	20	Amann et al. [23]
Lla	(23S, 271-289)[b]	CTATAATGCTTAAGTCCACG	20	Beimfohr et al. [24]
Btcl405	S-*-Btcl-0405-a-A-18	GAAGACCTTCATCGCTCA	10	Kurisu et al. [25]
Blen483	S-*-Blen-0483-a-A-18	CTTTCTGGTCAGGTACCG	5	Kurisu et al. [25]
Ecos1252	S-*-Ecos-1252-a-A-18	CTCGCGACTTCGCAACTC	5	Kurisu et al. [25]

[a] Oligonucleotide probe database [22].
[b] OPD name is not available and *E. coli* number of the target site is shown instead.

microscope (TCS-NT, Leica Microsystems, Heidelberg, Germany) equipped with Ar/Kr laser. Strong autofluorescence hindered specific detection when XRITC was the marker. Thus, the filter set for FITC was used to view the same image and these two images were superimposed by image processing. Autofluorescence was observed through both filter sets as it occurs over a wide range of fluorescent wavelengths [4]. However, probe specific fluorescence is visible only in XRITC image and therefore, it could be clearly distinguished in the merged image.

Fluorescent areas in the images were quantified with the Q600HR image analyzing software (Leica). The proportion of each specific probe group was represented by the ratio of the specific probe area to that of the eubacterial probe area in a given field. At least 5 different fields (more than 1000 cells) were examined for each sample and probe.

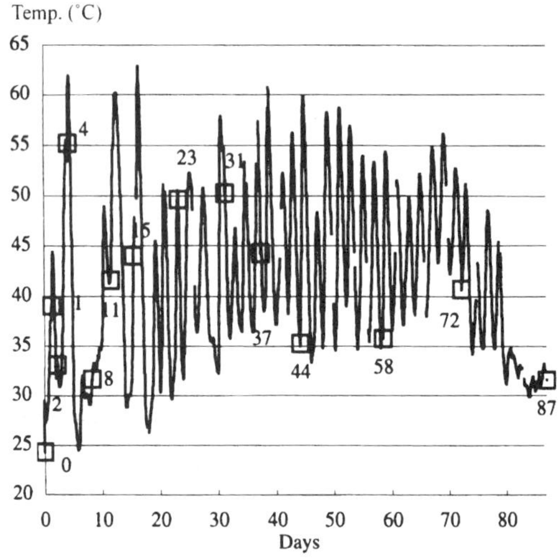

Fig. 2. Temperature inside the reactor. The day samples were taken is indicated in the figure.

2.8. EUB/DAPI ratio

The detection ability of cells by FISH was determined by hybridizing to bacterial probe EUB338 counterstaining with 4′,6-diamidino-2-phenylindole (DAPI). After the hybridization with EUB338, the slides were incubated in the dark with DAPI staining solution ($1\,\mu g\,ml^{-1}$ DAPI, 0.9 M NaCl and 20 mM Tris/HCl [pH 7.2]) for 5 min, washed with distilled water, and dried in air. The slides were mounted with antifade as described above. Subsequently, the samples were examined by an epifluorescent microscope (DMRBE, Leica) and color photomicrographs were taken. More than 2000 cells stained with DAPI were enumerated.

3. Results

3.1. Performance of the reactor

The temperature inside the reactor throughout the operation is shown in Fig. 2. The temperature reflects bacterial activity very well as it increases due to the heat generated by the decomposition of organic materials. The temperature started increasing after 2 days, and by day 4, it reached 65°C. From day 4 to day 6, the mixing of substratum stopped due to a rotor malfunction. As there was no mixing, aerobic degradation of the substrate almost ceased due to insufficient oxygen supply for the attached microorganisms. Thus, the temperature did not increase during this period. The temperature started increasing again after day 6, and it fluctuated periodically from day 30 to day 70. Fig. 3 shows a typical temperature profile at the steady state over a period of 48 h. The temperature increased just after the feeding, peaked at 50–60°C and then decreased to about 35°C. The carbon dioxide emission rate changed in a similar manner to that of the temperature, although the maximum rate occurred a couple of hours prior to the time of observed highest reactor temperature. After day 70, the peak reactor temperature decreased gradually and from day 80, no further temperature increase was observed.

A carbon balance was carried out on the system using the degradation ratio, which was defined as the ratio of CO_2-carbon discharged in 2 days to that of the organic carbon fed in a batch. It was 40–180% from day 10 to day 58 (Fig. 4). Degradation of accumulated organic

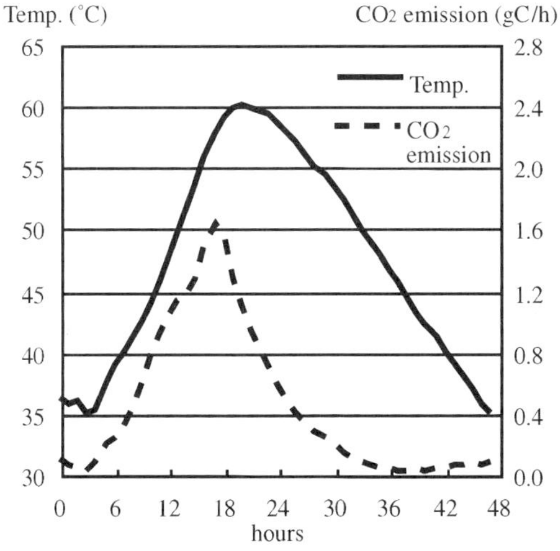

Fig. 3. Temperature and CO_2 emission after batch feed on day 44.

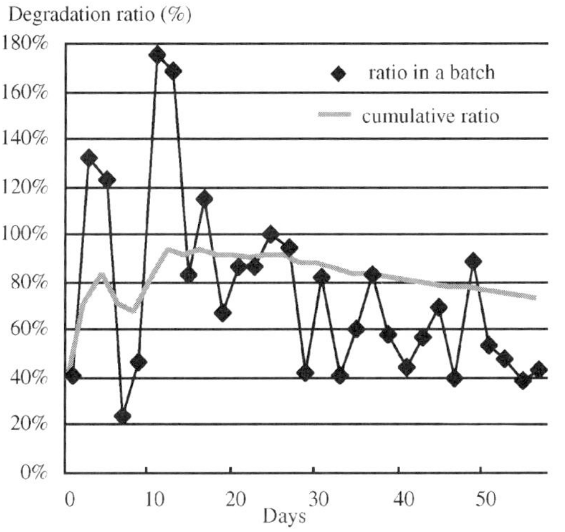

Fig. 4. Degradation ratio based on carbon balance.

matter in the reactor may have resulted in values over 100%. The cumulative degradation ratio showed that approximately 80% of total organic carbon input was converted to CO_2 and was discharged from the reactor in 58 days. Due to a faulty CO_2 meter, no data were available after day 58. However, considering the temperature fluctuations, the degradation ratio should have remained within the same range of value until day 70.

3.2. Quinone profile

Quinone profiles of the samples are shown in Fig. 5. Considerable differences in the profiles were observed from day 0 (inoculum) to day 1, and from day 1 to day

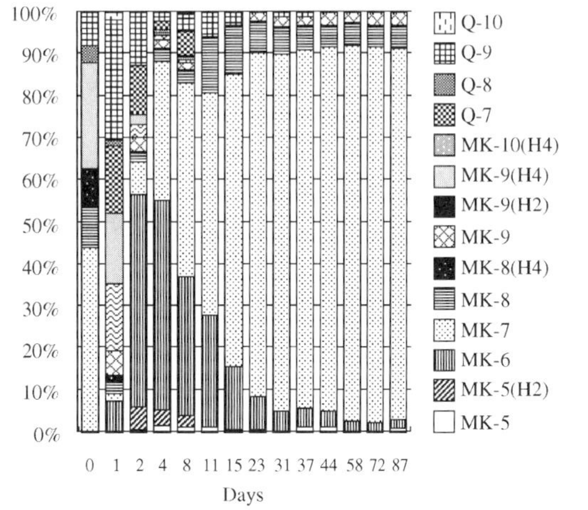

Fig. 5. Quinone profiles of samples taken from the reactor.

2. On day 2, MK-6 consisted of 50% of the total, whereas MK-7 was only 10% of the total. The proportion of MK-6 decreased after day 2 and MK-7 increased. On day 23 and later, MK-6 was less than 10%, MK-7 was more than 80%, and the remaining components were other menaquinones. No ubiquinones were detected.

According to Collins and Jones [3], only *Capnocytophaga, Desulfovibrio, Staphylococcus* and *Caryophanon* have MK-6 as a major quinone component. *Capnocytophaga, Desulfovibrio* are unlikely to predominate in a fully aerobic reactor as they are strict anaerobes. Thus, the detection of MK-6 may indicate the predominance of *Staphylococcus* and/or *Caryophanon*, which are facultative anaerobes. Bacteria that mainly possess MK-7 are *Cytophagales* and Gram positive bacteria with low G+C content (GPBLGC). FISH with probe CF319, specific to *Cytophaga/Flavobacterium* group [13], however, did not reveal the presence of this group (data not shown). Thus, bacterial with MK-7 were determined as GPBLGC. The bacteria community was dominated by GPBLGC from day 2 onwards, as *Staphylococcus* and *Caryophanon* are also members of GPBLGC. However, data of quinone patterns are available only for a limited number of bacterial species and hence other species might also be present in the community.

3.3. PCR-DGGE analysis

Fig. 6 shows the result of the PCR-DGGE analysis. The banding pattern of the inoculum was completely different from that of day 1 and subsequent samples. There was a remarkable change in the pattern on day 11; however, more consistent and stable patterns were obtained after day 23. The distinct banding patterns could be clearly identified by clustering analysis [9];

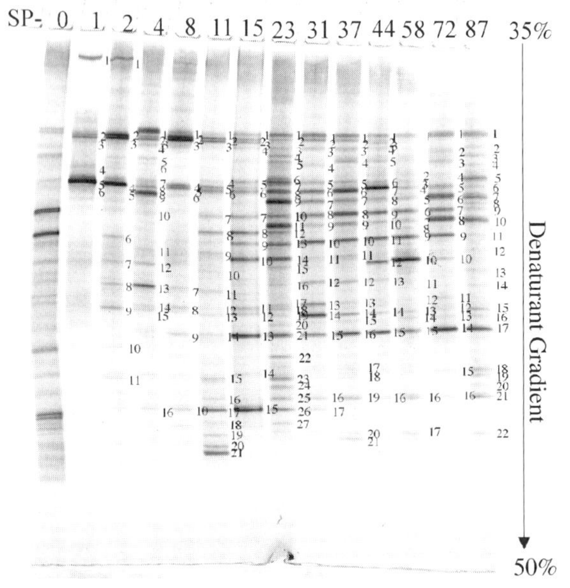

Fig. 6. A gel image of PCR-DGGE analysis. All the detectable bands were numbered and the numbers are shown on the right side of the bands. Samples are named SP0 (inoculum), SP1, SP2, etc. The bands are named as 'sample name'—'band no.', for example, the band no. 1 in the sample on day 4 is called 'SP4-1'.

accordingly, they could be classified into four clusters based on their similarity: (1) inoculum, (2) days 1–8, (3) days 11 and 15, and (4) days 23–87.

3.4. Sequencing of DGGE bands

Prominent DGGE bands from the samples of day 4, day 31 and some other distinctive bands, were sequenced to elucidate the affiliation of constituents in the community. Table 2 shows the results of homology search for the sequences obtained from the DGGE bands. Several sets of bands having identical migration positions were sequenced and all of them had identical sequences (Table 2). All 27 sequences had high similarity to only *Bacillaceae* and hence they could all be classified into *Bacillaceae*. This result collaborates the findings from the quinone profile analysis as *Bacillaceae* falls within the GPBLGC group.

3.5. FISH analysis

We designed species-specific oligonucleotide probes based on the sequences retrieved to further quantify their abundance in the samples (Table 3). Bacteria specific for probe Blen 483 were detected in all the samples tested. Their proportion increased and became more than 60% of the total eubacterial cells by day 15 and decreased to less than 10%. Thereafter, probe Btcl 405 hybridized to a substantial number of bacteria after day 4, although they were not major constituents of the bacterial community. Probe Lla specific cells were observed until day 15 and following day 44. The proportion of bacteria detected by Ecos 1252 decreased after day 2, and no further cells were detected after day 23.

4. Discussion

4.1. Methodology

The PCR-DGGE method reveals the community change at species level. The banding pattern of PCR-DGGE indicated the population dynamics very well in this study. Although the number of DGGE bands represents the complexity of the bacterial community, it does not necessarily represent the exact number of species in the sample. The DNA fragments of different sequences are not necessarily separated by DGGE [14]. As more than one copy of 16S rDNA exist in most of the bacteria [15] with different sequences among the copies, multiple bands may be obtained from a single strain. For example, we found that there are three base pair differences in the PCR target site (between 341 and 534 (*E. coli* position)) among ten copies of 16S rDNA of *B. subtilis* in the GeneBank database (data not shown).

In addition to monitoring of the community profile, PCR-DGGE was used to separate 16S rDNA sequences instead of the cloning technique. One advantage of the PCR-DGGE method over the cloning technique is that banding patterns from different samples can be compared and DNA fragments of interest can be selected for sequencing. The FISH method reconfirmed the existence of the species in the samples from which the sequences had been earlier retrieved by excising the DGGE bands. This reconfirmation cleared any doubts as to whether the retrieved sequences are chimeric PCR products. FISH further enabled the quantification of the species of interest, which helped clarify their importance in the community. The advantage of the quinone profile method is its quntitativeness, which was useful to verify the data obtained by the FISH method and the sequences retrieved by the DGGE method. Since a detailed and quantitative analysis is required to correlate the performance of a biological wastewater treatment process to its microbial community, the combination of these methods was shown to be a powerful means of achieving this objective.

Table 2
Homology of sequences of DGGE bands to described species

Band no.	Highest similarity[a]	Homology
SP4-1, SP31-1	*B. oleronius*	181/195
SP4-3, SP2-1, SP8-3	*Eco. saccharolyticus*	195/195
SP4-9	*Sac. thermophilus*	187/195
SP4-8, SP44-7	*Lcc. lactis* subsp. *lactis*	195/195
SP4-13	*B. badius*	188/194
SP11-5	*Mcc. carouselicus*	186/194
SP11-17	*B. fastidiosus*	186/194
SP15-10	*B. lentus*	189/194
SP23-11	*B. licheniformis*	192/195
SP31-2, SP8-2	*Staphyl. sciuri*	194/194
SP31-4	*B. licheniformis*	188/195
SP31-5	*B. lentus*	178/194
SP31-6	*B. licheniformis*	189/195
SP31-7	*B. thermocloacae*	187/195
SP31-8	*B. licheniformis*	190/195
SP31-10	*B. licheniformis*	185/195
SP31-12	*B. thermocloacae*	190/194
SP31-13	*B. thermocloacae*	188/194
SP31-14	*B. cohnii*	186/194
SP44-15	*B. pumilus*	183/194
SP58-10	*B. benzoevorans*	188/194

[a] *Lcc.*: *Lactococcus*, *B.*: *Bacillus*, *Eco.*: *Enterococcus*, *Sac.*: *Saccharococcus*, *Mcc.*: *Macrococcus*, *Staphyl.*: *Staphylococcus*.

4.2. Comprehension of the data for community analysis

Probe EUB338 hybridized to only 10–40% of the cells stained by DAPI (Table 3). The rest were considered to be inactive bacteria. Although the EUB/DAPI ratios were quite low, the percentages of probe-positive cells could be used for elucidation of the reactor performance, as only active biomass determines the treatment capability.

EUB338-negative cells were considered inactive based on the following rationale. All the cells visualized by DAPI were not large enough to be categorized as Eukaryotes. None of the cells in any samples were hybridized to probe Arch915, which is specific for Archaea [16]. These two observations indicated that almost all the cells belonged to the domain of bacteria. Daims et al. [17] reported that probe EUB338 failed to detect some bacteria that had more than two mismatches in the probe target site. However, GPBLGC, which dominated the community in the reactors, have none or only one mismatch in the target site except for a few species and therefore, EUB338 could have detected most of them. On the other hand, poor probe permeability into cells may result in a lower EUB to DAPI ratio. However, probe EUB338 hybridized very well with the pure cultures that were close relatives of the species identified by sequencing of DGGE bands. Thus,

Table 3
Detection and quantification of target bacteria by oligonucleotide probes and the ratio of EUB338-positive cells to DAPI-stained cells

Sample (SP)	1	2	4	8	11	15	23	31	37	44	58	72	87
Blen483	n.d.[a]	+[b,c]	7%[d]	n.d.	n.d.	62%	9%	6%	3%	+	+	n.d.	+
Btcl405	n.d.	–[e]	6%	n.d.	n.d.	5%	7%	14%	+	5%	14%	n.d.	3%
Lla	n.d.	5%	+	n.d.	n.d.	1%	–	–	–	17%	20%	n.d.	1%
Ecos1252	n.d.	11%	+	n.d.	n.d.	+	–	–	–	–	–	n.d.	–
EUB/DAPI	33%	19%	8%	19%	10%	17%	13%	14%	9%	41%	38%	22%	28%

[a] n.d.: no data.
[b] (+): detected.
[c] Ratio to the total bacteria was not measured if the ratio is negligible.
[d] Ratio to the total bacteria.
[e] (–): not detected.

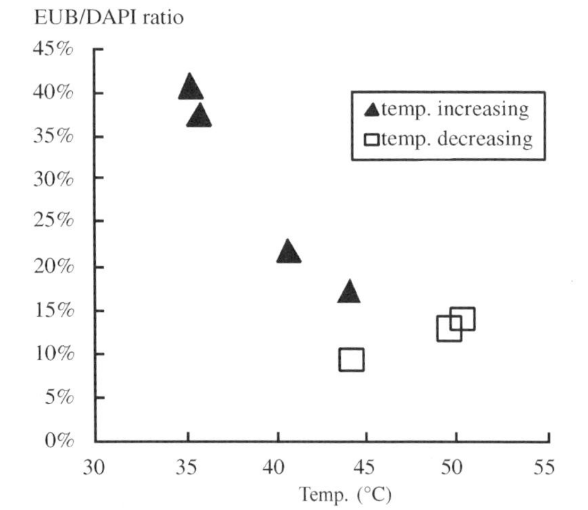

Fig. 7. The EUB/DAPI ratio plotted against the reactor temperature. The points were differentiated depending on the trend of the temperature change at the time of sampling.

the cells detected by DAPI but not by EUB probe could have been detected provided they contained sufficient number of ribosomes in their cells, that is, if they had been active. *Bacillus*, which were found to be the predominant bacterial group in the reactor, are also known to be spore formers in the presence of adverse conditions. In all probability, most of the inactive cells were spores of *Bacillus*. Their spherical shape and 1–2 μm size supports this supposition.

The community profiles obtained by the PCR-DGGE method and quinone profile were stable after day 23, even though samples were taken from the reactor at different stages (Fig. 2). As nucleic acids and quinones were extracted from both active and inactive cells, this result indicated that the constituent species of the total community remained unchanged after day 23. In contrast, the percentages derived by FISH fluctuated during the same period. This fluctuation may be due to the change in activity of the cells. Fig. 7 shows the temperature at the time of sampling and the EUB/DAPI ratio during the stable stage. This figure illustrates that the EUB/DAPI ratio, which is an indicator of the proportion of active cells, decreased with the time in a batch. Some species might become active immediately after feeding and become inactive after depletion of the substrate or other conditions, although the total community structure inclusive of inactive bacteria remained the same.

4.3. Process performance and community structure

The trends in temperature change can be separated into four periods: the initial period (day 0–10), the semi-stable period (day 10–30), the stable period (day 30–70), and the deterioration period (day 70–87). This separation is in good agreement with the distinct DGGE banding patterns. Thus, the relationship between the process performance and community structure could be discussed for the individual period in isolation.

During the initial period (from day 0 to day 10), some mesophilic fermentative bacteria like *Enterococcus*, *Lactococcus* and *Staphylococcus* were found. *Enterococcus* and *Lactococcus* can only degrade starch and sucrose fermentatively [18,19]. The fermentation of starch and sucrose in the substrate generates a small quantity of energy thus leading to a minute temperature increase. *Lactococcus* were also found in the samples after day 44. Since they were only found in the samples taken prior to the occurrence of high temperature (day 1, 2, 4, 8, 11, 15, 44, 58, 72 and 87), seeding by the substrate may have contributed to their presence and the high temperature would have eliminated them. The substrate was occasionally found to be contaminated at feeding in spite of the initial sterilization. Additionally, the abundance of MK-6 in this period implies the existence of a large number of *Staphylococcus sciuri*, as this is the only species that has MK-6 as a major quinone [3] among the identified DGGE bands.

From day 10 to day 30, the reactor performance was almost stable except for few batches. The quinone profile and the DGGE banding patterns showed that the community structure shifted to a stable state. On day 15, 62% of active cells were hybridized to probe Blen483. The known species among them are *B. lentus*, *B. firmus* and *B. badius*, which do not grow at or above 55°C [20]. Although the reactor temperature was above 55°C for several hours on day 12, just 3 days after, they not only survived but dominated the community. They probably had formed spores and survived when the temperature was above their survival limit, and germinated when the substrate was supplied at a favorable temperature for their growth. On the other hand, the only known target species of probe Btcl405, *B. thermocloacae* was detected by FISH on day 15. Although the temperature had been below their growth limit (40°C) [21] for more than a day (day 13–14), their detection by FISH implies that they were active on day 15. This phenomenon also indicates their tolerance for adverse conditions by spore formation.

The performance of the reactor and the community structure was stable from day 30 to day 70. Based on the sequencing of DGGE bands, the community was composed primarily of *Bacillus* spp., including the close relatives of *B. thermocloacae* (obligate thermophile), *B. licheniformis* (facultative thermophile) and *B. lentus* (mesophile). Relatives of *B. licheniformis* might be predominant because many DGGE bands were affiliated to them and target species of Blen483 and Btcl405 were minor components during this period. Facultative thermophiles have an advantage under the fluctuating temperature conditions as they can grow under both mesophilic and thermophilic conditions. *B. licheniformis* can utilize starch and sucrose [20] and therefore the predominant species might actually be *B. licheniformis* or a species with similar characteristics. Some obligate thermophilic and mesophilic *Bacillus* spp. would have also played some role in the reactor performance. The CO_2 emission rate stopped increasing temporarily at about 50°C (see Fig. 3) in most batches. This phenomenon can be attributed to the coexistence of these *Bacillus* spp. The substrate entered the reactor at mesophilic condition and mesophilic *Bacillus* spp. became active, generating heat. When the temperature increased up to 50°C, they became inactive causing the stagnation of CO_2 emission rate. At the same time, obligate thermophilic *Bacillus* spp. became active, generating CO_2, leading to an increasing emission rate.

Even after the performance deteriorated and the temperature ceased to increase after substrate addition (day 80), the community structure seemed to remain the same as before according to the DGGE result, although some of the bands on day 87 may be derived from dead cells. Thus, this deterioration was not probably caused by community change, but by some physical conditions. A probable cause could be the moisture content in the wood chips. It was 50–65% from day 10 to day 60 but increased above 70% after day 70 (data not shown). High moisture content reduces the diffusion rate of oxygen thereby hindering active aerobic metabolism. Another reason may be the accumulation of inorganic material. Substances that cannot be gasified remain in the reactor in this process. The high concentration of inorganic material may have inhibited bacterial growth.

4.4. Strategy to maintain stable good performance

The most important factor in this process is the temperature increase. If the temperature is not increased adequately, water gets accumulated in the reactor, and continuous operation becomes impossible. Inoculation of appropriate bacteria is one of the promising ways to ensure temperature increase at the beginning of the operation. According to the results, the mixture of the mesophilic, facultative thermophilic and obligate thermophilic *Bacillus*, namely *B. lentus*, *B. licheniformis* and *B. thermocloacae*, can be good inocula. The interval between feeding might also be important to ensure the temperature increase if those bacteria become active in turn. The lack of increase in temperature after 80 days of operation was not directly caused by the community change. It is necessary to improve the physico-chemical conditions of the substratum, rather than to manipulate microbial community, to overcome this problem.

5. Conclusions

In the present study, microbial community structure and its dynamics was determined quite well using a combination of PCR-DGGE, FISH and quinone profile method. The performance of the reactor could be explained well by the community structure. To characterize and comprehend biological wastewater treatment processes by community structure, this combination of methods is a powerful tool. The community structure of TCOP during the period of good degradation shows the coexistence of mesophilic, facultatively thermophilic and obligate thermophilic *Bacillus* spp., such as *B. lentus*, *B. licheniformis* and *B. thermocloacae*, respectively. They may act in turns depending on the conditions. The in situ physiology of identified species needs further elucidation.

References

[1] Wagner M, Amann R, Lemmer H, Schleifer K-H. Probing activated sludge with oligonucleotides specific for proteobacteria: inadequacy of culture-dependent methods for describing microbial community structure. Appl Environ Microbiol 1993;59(5):1520–5.

[2] Hiraishi A. Respiratory quinone profiles as tools for identifying different bacterial populations in activated sludge. J Gen Appl Microbiol 1988;34:39–56.

[3] Collins MD, Jones D. Distribution of isoprenoid quinone structural types in bacteria and their taxonomic implications. Microbiol Rev 1981;45(2):316–54.

[4] Amann RI, Ludwig W, Schleifer K-H. Phylogenetic identification and in situ detection of individual microbial cells without cultivation. Microbiol Rev 1995;59(1):143–69.

[5] Muyzer G, Waal EC, Uitterlinden AG. Profiling of complex microbiol populations by denaturing gradient gel electrophoresis analysis of polymerase chain reaction-amplified genes coding for 16S rRNA. Appl Environ Microbiol 1993;59(3):695–700.

[6] Rolleke S, Muyzer G, Wawer C, Wanner G, Lubitz W. Identification of bacteria in a biodegraded wall painting by denaturing gradient gel electrophoresis of PCR-amplified gene fragments coding for 16S rRNA. Appl Environ Microbiol 1996;62(6):2059–65.

[7] Fujie K, Hu HY, Tanaka H, Urano K. Ecological studies of aerobic submerged biofilter on the basis of respiratory quinone profiles. Water Sci Tech 1994;29(7):373–6.

[8] Zhu H, Qu F, Zhu LH. Isolation of genomic DNAs from plants, fungi and bacteria using benzyl chloride. Nucleic Acids Res 1993;21(22):5279–80.

[9] Kurisu F, Satoh H, Mino T, Mastuo T. Microbial community analysis of thermophilic contact oxidation process by using PCR-DGGE method. In: Matsuo T, Hanaki K, Takizawa S, Sato H, editors. Advances in water and wastewater treatment technology —molecular technology, nutrient removal, sludge reduction, and environ-

mental health. Netherlands: Elsevier Science. 2000, in press.

[10] Amann RI. In situ identification of micro-organisms by whole cell hybridization with rRNA-targeted nucleic acid probes. In: Akkermans ADI, van Elsas JD, de Bruijn FJ, editors. Molecular microbial ecology manual. Dordrecht, The Netherlands: Kluwer Academic Publishers, 1995. p. 3.3.6.1.–15.

[11] Manz W, Wagner M, Amann R, Schleifer K-H. In situ characterization of the microbiol consortia active in two wastewater treatment plants. Water Res 1994;28(8):1715–23.

[12] Stahl DA, Amann R. Development and application of nucleic acid probes. In: Stackbrandt E, Goodfellow M, editors. Nucleic acid techniques in bacterial systematics. Chichester, England: Wiley, 1991. p. 205–48.

[13] Manz W, Amann R, Ludwig W, Vancanneyt M, Shleifer K-H. Application of a suite of 16S rRNA-specific oligonucleotide probes designed to investigate bacteria of the phylum Cytophaga–Flavobacter–Bacteroides in the natural environment. Microbiology 1996;142:1097–106.

[14] Cotton RGH. Current methods of mutation detection. Mutation Res 1993;285:125–44.

[15] Fogel GB, Collins CR, Li J, Brunk CF. Prokaryotic genome size and SSU rDNA copy number: estimation of microbial relative abundance from a mixed population. Microb Ecol 1999;38(2):93–113.

[16] Burggraf S, Mayer T, Amann R, Schadhauser S, Woese CR, Stetter KO. Identifying members of the domain Archaea with rRNA-targeted olignucleotide probes. Appl Environ Microbiol 1994;60(9):3112–9.

[17] Daims H, Bruhl A, Amann R, Schleifer K-H, Wagner M. The domain-specific probe EUB338 is insufficient for the detection of all bacteria: development and evaluation of a more comprehensive probe set. System Appl Microbiol 1999;22(3):434–44.

[18] Farrow JAE, Kruze J, Phillips BA, Bramley AJ, Collins MD. Taxonomic studies on *Streptococcus bovis* and *Streptococcus equinus*: description of *Streptococcus alacotolyticus* sp. nov. and *Streptococcus saccharolyticus* sp. nov. System Appl Microbiol 1984;5(2):467–82.

[19] Shleifer KH, Kraus J, Dvorak C, Kilpper-Balz R, Collins MD, Fisher W. Transfer of *Streptococcus lacits* and related *Streptococci* to the genus *Lactococcus gen.* nov. System Appl Microbiol 1985;6:183–95.

[20] Claus D, Berkeley RCW. Endospore-forming Gram-positive rods and cocci: *Bacillus*. In: Sneath PA, Mair NS, Sharpe ME, Holt JG, editors. Bergy's manual of systematic bacteriology, vol. 2, 2nd ed. Baltimore, MD, USA: Williams & Wilkins, 1986. p. 1105–38.

[21] Demharter W, Hensel R. *Bacillus thermocloacae* sp., nov., a new thermophilic species from sewage sludge. System Appl Microbiol 1989;11(3):272–6.

[22] Alm EW, Oerther DB, Larsen N, Stahl DA, Raskin L. The oligonucleotide probe database. Appl Environ Microbiol 1996;62(10):3557–9.

[23] Amann RI, Krumholz L, Stahl DA. Fluorescent-oligonucleotide probing of whole cells for determinative, phylogenetic, and environmental studies in microbiology. J Bacteriol 1990;172(2):762–70.

[24] Beimfohr C, Krause A, Amann R, Ludwig W, Schleifer K-H. In situ identification of *Lactococci*, *Enterococci* and *Streptococci*. System Appl Microbiol 1993;16(3):450–6.

[25] Kurisu F, Satoh H, Mino T, Matsuo T. Characterization of bacterial community structures in thermophilic aerobic wastewater treatment processes by ribosomal RNA approaches. Appl Environ Microbiol, 2000, submitted for publication.

PERGAMON

Water Research 36 (2002) 439–444

WATER RESEARCH

www.elsevier.com/locate/watres

Active biomass in activated sludge mixed liquor

G.L. Cronje, A.O. Beeharry, M.C. Wentzel*, G.A. Ekama

Water Research Group, Department of Civil Engineering, University of Cape Town, Rondebosch 7701, South Africa

Received 20 June 2000; received in revised form 5 January 2001; accepted 22 January 2001

Abstract

The engineering and technology of the activated sludge system are reasonably well established, with systems implemented worldwide for the biological removal of C, N and/or P. Parallel to this development, significant advances have been made in the microbiological and biochemical areas of activated sludge. These advances have been driven by the development of new analytical techniques that allow microorganisms to be studied in situ in the activated sludge environment. However, there has been little cross-linking and overlap between the engineering and technology and microbiology and biochemistry paradigms. In particular, the information from the microbiology and biochemistry has not been integrated into the engineering and technology paradigm, to enable improved system design and optimization. One area that can form a starting point to build bridges between the two paradigm sets, is measurement of active biomass. The current design and simulation models invariably include active biomass for each organism group as fundamental parameters which define quantitatively the kinetic rates of the relevant biological processes. However, these parameters remain purely hypothetical because to date they have not been quantitatively measured; their acceptance is based on the consistency of model predicted results over a wide range of application. This paper describes developments in quantitative measurement of the heterotrophic and autotrophic active biomass concentrations within the engineering and technology paradigm, and the formulation of a multinational project which will attempt to link these measurements and the defined engineering environment to the new microbiological and biochemical analytical techniques. It is hoped that this project will facilitate integration of the two paradigms sets.

Keywords: Activated sludge system; Active biomass; Design; Heterotrophic; Autotrophic; Models; Microbiology; Biochemistry

1. Introduction

To comply with more stringent effluent legislation, over the past two decades significant advances have been made in the areas of engineering (design) and technology (implementation and operation) of the single sludge activated sludge system. Activated sludge systems have been successfully designed and implemented at full-scale to progressively include the biological removal of carbon (C), nitrogen (N) and phosphorus (P). This implementation has been aided by the development of a suite of steady state design models (e.g. [1–3]) and kinetic simulation models (e.g. [4–9]). These models are based on a common conceptualization of the processes acting in the system. In terms of this group of models, in the bioreactor of the non-nitrifying aerobic activated sludge system, the mixed liquor organic (volatile) suspended solids (MLOSS) is made up of three components: (1) ordinary heterotrophic organism (OHO) active biomass, (2) endogenous residue and (3) inert material. In the nitrifying aerobic and anoxic/aerobic activated sludge systems, a fourth component is included; (4) autotrophic organism (AO) active biomass. All four MLOSS components settle out in the secondary settling tank and are returned to the bioreactor via the underflow recycle; these components leave the system via the waste flow. If an anaerobic reactor is included to stimulate biological excess phosphorus removal (BEPR),

*Corresponding author. Tel.: +27-21-650-2583; fax: +27-21-689-7471.

E-mail address: markw@eng.uct.ac.za (M.C. Wentzel).

PII: S0043-1354(01)00226-3

additionally (5) phosphate accumulating organism (PAO) active biomass and (6) this organism group's endogenous residue will contribute to the MLOSS [7,9]. The active biomass components of the MLOSS mediate the relevant biological processes deemed to be of importance; OHO's COD removal and denitrification, AO's nitrification and PAO's BEPR and COD removal.

Historically the MLOSS has been measured as a lumped parameter, via the VSS or COD test [10]. Specific rates for the biological processes (e.g. denitrification; oxygen utilization) often were (and still are) expressed in terms of this lumped parameter. However, from the above, only parts of the MLOSS are active biomasses, and only these parts mediate the relevant biological processes, e.g. OHOs for COD removal and denitrification. Accordingly, the specific rates for the relevant (and associated) biological processes should be expressed in terms of the appropriate active biomass concentration to allow a meaningful comparison of rates measured in different systems. More recently, with the proliferation of kinetic simulation computer programmes that invariably include active biomass concentrations as parameters (e.g. Biowin, Simba, GPX, UCTOLD, UCTPHO), these parameters and the use of specific rates in terms of them, have become much more widely accepted. However, this acceptance has not been driven by sound scientific proof of the active biomass concept, but rather by the convenience of the computer programmes. It must be remembered that active biomass exists only hypothetically within the structure of the design procedures and kinetic models. Although indirect evidence does provide some support for the active biomass parameters (by consistency between observations and predictions over a wide range of conditions, e.g. [4,5,8]), these have not been directly measured experimentally and compared to the hypothetical model values. This deficiency casts a measure of uncertainty on the entire framework within which the models have been developed and is a major weakness in the models. The problem in measurement has been the lack of suitable experimental techniques. Recently a simple batch test procedure has been developed to quantify OHO active biomass concentration, based on the concepts incorporated in the models [11,12]. Evaluation of this batch test method, and comparison of the OHO active biomass concentrations derived from the batch test with theoretical concentrations from the models has had mixed results [13]. Thus, uncertainty around the active biomass concept largely remains.

Parallel to the developments in the engineering and technology of the activated sludge system, significant advances have been made in the microbiological and biochemical areas. As researchers in these fields have moved away from pure culture work to the activated sludge environment, a number of new analytical techniques have been developed to study microorganisms insitu, e.g. ATP analysis [14], DNA analysis [15], quinone profiling [16], microautoradiography [17], using florescent probes for ribosomal RNA [18,19]. While the microbiological and biochemical knowledge and developments have made a considerable contribution to the understanding of the biological nutrient removal activated sludge system, the full potential of these developments have yet to be realised for the system. It remains for the results that these techniques provide to be integrated with the design and kinetic modelling theory. The consequence of this is that the engineering and technology (modelling) paradigm has largely worked independently of the microbiological and biochemical paradigm. To facilitate links and overlap between the two paradigm sets, the new developments in the microbiological and biochemical analytical techniques can be implemented to address the deficiency in the engineering and technology paradigm of the active biomass concept. This should prove possible because, in contrast to the more traditional analytical techniques, the new techniques provide quantitative information, a prerequisite for modelling. Some initial integration between modelling and these techniques has been started (e.g. [20,21]), but this is still in its infancy.

This paper describes developments in (i) measuring the heterotrophic and autotrophic active biomass concentrations within the engineering and technology paradigms, and (ii) formulating a multinational project which will attempt to link these measurements and the defined engineering environment to the new microbiological and biochemical analytical techniques.

2. Measurement of heterotrophic active biomass

Ubisi et al. [22,23] describe the development of a simple batch test to quantify the OHO active biomass concentration, that is based on the concepts incorporated in the models for activated sludge systems. In this test a small sample of mixed liquor is drawn from the activated sludge system and mixed with raw wastewater in a batch reactor and the oxygen utilization rate (OUR) and nitrate and nitrite concentrations monitored with time. In parallel, a similar batch test is conducted on the raw wastewater without mixed liquor addition. From the analysis of the OUR and nitrate and nitrite responses of the two parallel tests, the mixed liquor OHO active biomass concentration can be quantified. Wentzel et al. [13] evaluated this batch test method by drawing mixed liquor samples from a well defined laboratory-scale anoxic/aerobic activated sludge system operated at 12 and 20 days sludge age. They compared the results from the batch tests with theoretical values for OHO active biomass concentrations from steady state design [1] and kinetic simulation [5] models, see Fig. 1a and b. From the comparison they concluded that the results obtained were both encouraging and perplexing. With the parent system at 12 days sludge age (Fig. 1a), the agreement between measured and theoretical values was remarkably good. However, with the parent system at 20 days sludge age (Fig. 1b) the agreement was poor, with the theoretical values being about 2 times those measured. Wentzel et al. could provide no explanation for this inconsistency, but concluded that the results do indicate that the batch test method may prove to be a valuable tool that can be used to provide greater insight into the behaviour of the aerobic and anoxic/aerobic activated sludge systems.

In a subsequent investigation, the experimental work of Wentzel et al. has been repeated at the University of Cape Town. The same system was operated, but at a sludge age of 10 days. The batch tests were repeated, and a poor correlation between measured and theoretical OHO active biomass concentrations was obtained (see Fig. 2a), similar to that obtained by Wentzel et al. [13] at the 20 days sludge age (see Fig. 1b). In examining the response pattern, it was observed that the OUR response of the OHO active biomass added to the batch reactor with the raw wastewater partially masks the response of the OHO active biomass from the mixed liquor. Accordingly, a potential source of error in the test arises when subtracting the wastewater OHO active biomass concentration (determined from the OUR for the wastewater only) from that for the mixed liquor + wastewater OHO active biomass concentration, to derive the mixed liquor OHO active biomass concentration. To eliminate this as a source of error, the batch test was modified to remove the OHO active biomass from the wastewater in the batch test. This was done by flocculating and filtering the raw wastewater; 10 ml of stock aluminium sulphate $[Al(SO_4)_3 \cdot 15H_2O$, stock at $50\,g\,l^{-1}]$ were added per l wastewater, the mixture was

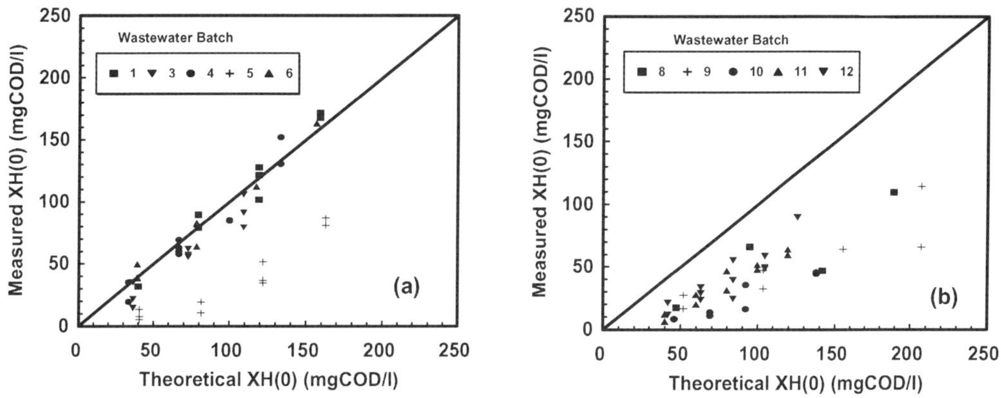

Fig. 1. Measured vs. theoretical heterotrophic active biomass concentration at time 0 [XH(0)] in the batch test due to addition of mixed liquor drawn from the parent system at (a) 12 days and (b) 20 days sludge age. Each data series represents batch tests conducted when the parent system received a single batch of wastewater as influent.

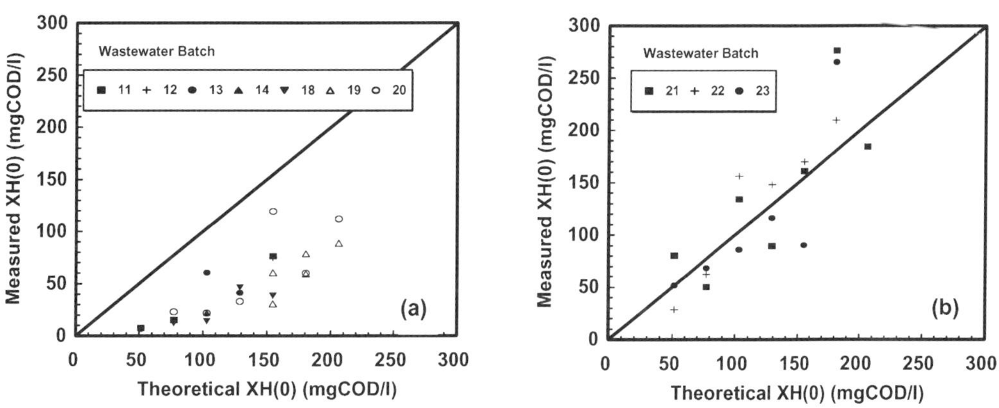

Fig. 2. Measured vs. theoretical heterotrophic active biomass concentration at time 0 [XH(0)] in the batch test due to addition of mixed liquor drawn from the parent system at 10 days sludge age with (a) raw wastewater addition and (b) flocculated filtered wastewater addition. Each data series represents batch tests conducted when the parent system received a single batch of wastewater as influent.

stirred rapidly (~200 rpm) for 2 min (rapid mix phase) and then slowly (~1 rpm) for 30 min (flocculation and settling phase). A clear liquid supernatant layer developed and this was drawn off and filtered through a glass fibre filter (Whatman's GF/C). Aerobic batch tests on the flocculated filtered wastewater only indicated that the procedure effectively removed all active biomass from the wastewater—no measurable OUR was observed. The mixed liquor drawn from the parent system was then added to flocculated filtered wastewater, and the batch test for measuring OHO active biomass concentration implemented, following the procedure set out in [13,22,23]. With this modification the OHO active biomass in the wastewater is eliminated from the batch test and no longer contributes to the OUR, and accordingly does not have to be subtracted to give the mixed liquor OHO active biomass concentration added from the parent system. A number of batch tests were conducted, and measured and theoretical OHO active biomass concentrations are compared in Fig. 2b; good correlation was obtained. The modification to the batch test appears successful and the investigation is continuing.

3. Measurement of autotrophic active biomass

Following the success of measuring the OHO active biomass concentration, a preliminary investigation has been started to determine whether a similar batch test can be developed to quantify AO active biomass concentration. In the batch test a small quantity of mixed liquor drawn from the parent system is mixed with activated sludge system effluent supplemented with ammonium chloride (50 mg NH_4^+ l^{-1} batch volume), the AO substrate. The nitrate and nitrite concentrations with time are then monitored. From the increase in the nitrate and nitrite concentrations with time, the AO active biomass concentration at the start of the test can be determined [24]. (A similar test has been applied by Orhon et al. [25] to measure the AO maximum specific growth rate, with some success.) Measured versus theoretical AO active biomass concentrations from a set of nine such batch tests are shown in Fig 3. Although the values are of similar magnitude, the theoretical values are approximately $2\frac{1}{2}$ times the measured values. However, it does appear that this test method holds promise, and will be investigated further as part of a proposed joint project, see below.

4. Linking the engineering and technology and microbiology and biochemistry paradigms

Although the developments above constitute a significant advance in attempting to resolve a major weakness within the steady state design and kinetic simulation models for activated sludge systems proposed within the engineering and technology paradigm, namely the lack of *independent* quantification of active biomass concentrations, the interpretation and analysis of the data remains firmly rooted within this paradigm. No cross-linking exists with the microbiological and biochemical paradigm. To overcome this shortfall, a project is being initiated in which both engineering (Dept. Civil Eng., Univ. of Cape Town, South Africa) and microbiological (Dept. Microbiol., Univ. of Pretoria, South Africa; Centre for Water and Wastewater Research, Tech. Natal, South Africa; Advanced Wastewater Management Centre, Univ. Queensland, Australia) based research groups will participate. Funding for the project has been obtained from Water Research Commission (South Africa) and Lyonnaise Des Eaux (France), and an application is pending from Cooperative Research Centre (CRC) for Waste Management and Pollution Control (Australia). In this project, activated sludge systems will be operated at laboratory scale for which the engineering parameters are closely controlled and defined. The systems will be monitored and all relevant performance parameters measured, and used to make the usual consistency checks, such as COD and N mass balances. Samples will be drawn from these systems and the engineering and technology type tests detailed above will be conducted to determine OHO and AO active biomass concentrations, and the measured values compared to model theoretical values. Additionally, the samples will be subjected to a range of the newly developed microbiological and biochemical analytical methods, in particular FISH, RNA analysis, DAPI staining, and ATP analysis. This information will be used to correlate the results from the different analytical

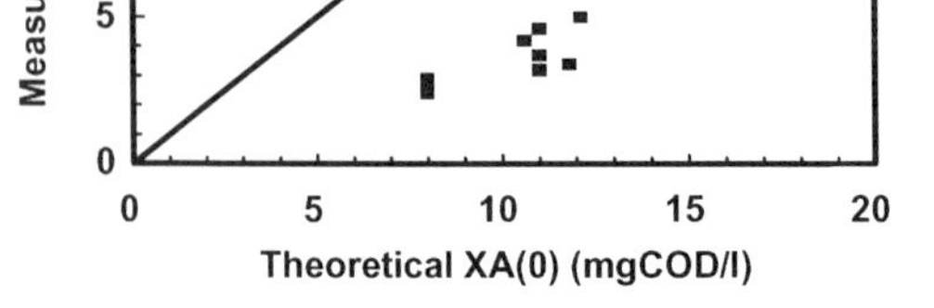

Fig. 3. Measured vs. theoretical autotrophic active biomass concentration at time 0 [XA(0)] in the batch test due to addition of mixed liquor drawn from the parent system at 20 days sludge age to effluent supplemented with ammonium chloride.

procedures, and hence find ways to integrate the two approaches.

5. Conclusions

Although significant developments have taken place in both the engineering and technology and the microbiology and biochemistry areas of the activated sludge system for treating mainly domestic wastewaters, these have proceeded on two parallel, but separate paths. Within the engineering and technology, the activated sludge system has become well established, with systems implemented worldwide for the biological removal of C, N and/or P. This implementation has been aided by the development of a suite of steady state and kinetic simulation models which have facilitated optimization of design and operation. Parallel to this development, significant advances have been made in the microbiological and biochemical areas of activated sludge. These advances have been driven by the development of new analytical techniques that allow microorganisms to be studied insitu in the activated sludge environment. However, there has been little cross-linking and overlap between the engineering and technology and microbiology and biochemistry paradigms. In particular, the information from the microbiology and biochemistry has not been integrated into the engineering and technology paradigm, to enable improved design and optimization. One area that can form a starting point to build bridges between the two paradigm sets is active biomass. The current design and simulation models invariably include active biomasses as fundamental parameters (OHO, AO and PAO), yet these parameters remain hypothetical as they have not been measured and favourably compared to theoretical values. Recently, new test methods have been developed to quantify OHO and AO active biomass concentrations. However, the interpretation and analysis of the data from these tests remains firmly rooted within the engineering and technology paradigm. No cross-linking exists with the microbiological and biochemical paradigms. To overcome this shortfall, a multinational collaborative project is being initiated which will attempt to link these measurements and the defined engineering environment to the new microbiological and biochemical analytical techniques. It is hoped that this project will facilitate the development of links and overlap between the two paradigms.

Acknowledgements

This research was supported jointly by the Water Research Commission and the National Research Foundation and is published with their permission.

References

[1] WRC. Theory, design and operation of nutrient removal activated sludge processes. Water Research Commission, P.O. Box 824, Pretoria 0001, South Africa, 1984.

[2] Wentzel MC, Dold PL, Ekama GA, Marais GvR. Biological excess phosphorus removal—steady state process design. Water SA 1990;16(1):29–48.

[3] Maurer M, Gujer W. Prediction of the performance of enhanced biological phosphorus removal plants. Water Sci Tech 1994;30(6):333–44.

[4] Dold PL, Ekama GA, Marais GvR. A general model for the activated sludge process. Prog Water Technol 1980;12:47–77.

[5] Dold PL, Wentzel MC, Billing AE, Ekama GA, Marais GvR. Activated sludge simulation programs. Water Research Commission, P.O. Box 824, Pretoria 0001, South Africa, 1991.

[6] Henze M, Grady Jr. CPL, Gujer W, Marais GvR, Matsuo T. Activated sludge model No. 1. IAWQ Scientific and Technical Report No. 1, IAWQ, London, 1987.

[7] Henze M, Gujer W, Mino T, Matsuo T, Wentzel MC, and Marais GvR. Activated sludge model No. 2. IAWQ Scientific and Technical Report No. 3, IAWQ, London, 1995.

[8] Van Haandel APC, Ekama GA, Marais GvR. The activated sludge process 3—single sludge denitrification. Water Res 1981;15:1135–52.

[9] Wentzel MC, Ekama GA, Marais GvR. Processes and modelling of nitrification denitrification biological excess phosphorus removal systems. Water Sci Tech 1992;25(6): 59–82.

[10] Standard Methods. Standard methods for the examination of water and wastewater. 16th ed. American Public Health Association, 1015 15th Street NW, Washington, DC 200005, USA, 1985.

[11] Kappeler J, Gujer W. Estimation of kinetic parameters of heterotrophic biomass under aerobic conditions and characterization wastewater for activated sludge modelling. Water Sci Tech 1992;25(6):125–40.

[12] Wentzel MC, Mbewe A, Ekama GA. Batch test for measurement of readily biodegradable COD and active organism concentrations in municipal wastewaters. Water SA 1995;21(2):117–24.

[13] Wentzel MC, Ubisi MF, Ekama GA. Heterotrophic active biomass component of activated sludge mixed liquor. Water Sci Tech 1998;37(4-5):79–87.

[14] Nelson PO, Lawrence AW. Microbial viability measurements and activated sludge kinetics. Water Res 1980;14:217–25.

[15] Liebeskind M, Dohmann M. Improved method of activated sludge biomass determination. Water Sci Tech 1994;29(7):7–13.

[16] Hu HY, Nozawa M, Fujie K, Makabe T, Urano K. Analysis of microbial acclimation to refractory chemicals in wastewater using respiratory quinone profiles. Water Sci Tech 1998;37(4-5):407–11.

[17] Nielsen PH, Anreasen K, Wagner M, Blackall LL, Lemmer H, Seviour RJ. Variability of type 021N in activated sludge as determined by in situ identification and

population monitoring in activated sludges. Water Sci Tech 1998;37(4-5):423–30.

[18] Wagner M, Amman RI, Kampfer P, Assmus B, Hartmann A, Hutzler P, Springer N, Schleifer KH. Identification and insitu detection of Gram-negative filamentous bacteria in activated sludge. System Appl Microbiol 1994;17:405–17.

[19] Water Science Technology. Microorganisms in activated sludge and biofilm processes. Water Sci Tech 1998; 37(4–5).

[20] Urbain V, Mobarry B, De Silva V, Stahl A, Rittmann RE, Manem J. Integration of performance, molecular biology and modeling to describe the activated sludge process. Water Sci Tech 1998;37(4-5):223–9.

[21] Wagner M, Noguera DR, Juretschko S, Rath G, Koops HP, Schleifer KH. Combining flourescent in situ hybridization (FISH) with cultivation and mathematical modeling to study population structure and function of ammonia-oxidizing bacteria in activated sludge. Water Sci Tech 1998;37(4–5):441–9.

[22] Ubisi MF, Wentzel MC, Ekama GA. Activated sludge mixed liquor heterotrophic active biomass. Water SA 1997;23(3):239–48.

[23] Ubisi MF, Wentzel MC, Ekama GA. Organic and inorganic components of activated sludge mixed liquor. Research Report W94, Dept. Civil Eng., Univ. of Cape Town, Rondebosch 7700, South Africa, 1997.

[24] Beeharry AO. Measurement of the nitrifier maximum specific growth rate. BSc undergraduate thesis, Dept. Civil Eng., Univ. of Cape Town, Rondebosch 7701, South Africa, 1999.

[25] Orhon D, Genceli EA, Sözen S. Experimental evaluation of the nitrification kinetics for tannery wastewaters. Water SA 2000;26(1):43–50.

PERGAMON

Water Research 36 (2002) 445–459

WATER RESEARCH

www.elsevier.com/locate/watres

Role of filamentous microorganisms in activated sludge foaming: relationship of mycolata levels to foaming initiation and stability

Francis L. de los Reyes III[1], Lutgarde Raskin*

Department of Civil and Environmental Engineering, University of Illinois at Urbana-Champaign, 3221 Newmark Civil Engineering Laboratory, 205 North Mathews Avenue, Urbana, IL 61801, USA

Received 17 July 2000; received in revised form 17 January 2001; accepted 25 January 2001

Abstract

The relationship between the levels of mycolic acid-containing actinomycetes (mycolata), *Gordonia* spp. and *Gordonia amarae*, and foam initiation and stability was characterized using: (1) batch tests involving addition of *G. amarae* cells to activated sludge, (2) analysis of a full-scale activated sludge plant that experienced seasonal foaming, and (3) a study of lab-scale activated sludge reactors augmented with *G. amarae*. Using batch tests, threshold *Gordonia* levels for foam formation and foam stability were determined to be approximately $2 \times 10^8\,\mu m\,ml^{-1}$ and $1 \times 10^9\,\mu m\,ml^{-1}$, respectively. In the full-scale plant, the levels of *Gordonia* spp. and *G. amarae* increased during the course of foaming, and the foam formation threshold of $2 \times 10^8\,\mu m\,ml^{-1}$ corresponded to the onset of foaming. This value was also verified in lab-scale reactor washout experiments, where decreasing mycolata levels were observed during the course of foam dissipation. The foam stability threshold of $1 \times 10^9\,\mu m\,ml^{-1}$ was verified in lab-scale reactor studies. The increase in the levels of *Gordonia* spp. and *G. amarae* in the full-scale plant corresponded to an increase in temperature, suggesting that the growth of *Gordonia* spp. was favored during warmer periods. © 2002 Elsevier Science Ltd. All rights reserved.

Keywords: Activated sludge; Filamentous foaming; Nocardioforms; Foam threshold; *Gordonia amarae*; 16S ribosomal RNA

1. Introduction

Foam on the surfaces of activated sludge aeration basins and secondary clarifiers can be classified as permanent (a thick, brown foam is always present), or temporal (seasonal, or even sporadic). In both cases, the mycolic acid-containing actinomycetes, or mycolata (particularly *Gordonia amarae*, formerly *Nocardia amarae*) [1,2], and "*Candidatus* Microthrix parvicella" (hereinafter referred to as "*M. parvicella*") are believed to be the main causative organisms of filamentous foaming [3,4]. The transient nature of foam occurrence may therefore be analyzed by monitoring changes in the levels of these microorganisms over time. This approach has been used in previous studies [4–6], in which the unique morphology and staining characteristics of some mycolata (the nocardioforms) were used to quantify their levels in activated sludge.

Despite these efforts, the appearance and disappearance of foam in many activated sludge plants is not completely understood. This may in part be due to the limitations of staining and microscopy techniques in identifying and quantifying causative organisms. Without a clear link between foaming and levels of causative organisms, contradictory factors in foaming causes (and eventual cures) cannot be resolved [6,7]. Furthermore, the diversity of filaments in foam and identification

*Corresponding author. Tel.: +1-217-333-6964; fax: +1-217-333-6968.

E-mail address: lraskin@uiuc.edu (L. Raskin).

[1] Present address: Department of Civil Engineering, North Carolina State University, Campus Box 7908, Raleigh, NC 27695, USA.

0043-1354/02/$ - see front matter
PII: S0043-1354(01)00227-5

difficulties may have contributed to the lack of understanding. To help resolve these issues, population shifts of putative foam-causing organisms before, during, and after foaming need to be quantified.

To link foaming to levels of foam-causing organisms, group-, genus-, and species-specific small-subunit (SSU) ribosomal RNA (rRNA)-targeted oligonucleotide probes for the mycolata, *Gordonia* spp. and *Gordonia amarae*, respectively, were developed [8,9]. Probes have also been developed for another putative foam-former, "*M. parvicella*" [10]. We previously used membrane hybridizations to quantify relative rRNA levels in activated sludge and developed a quantitative fluorescence in situ hybridization (FISH) method for estimating the biomass associated with *Gordonia* spp. [11]. These two hybridization formats were subsequently used to follow shifts in population levels of *Gordonia* spp. and "*M. parvicella*" before, during, and after a seasonal foaming incident in a full-scale plant [12]. The results showed that the relative levels of rRNA and volatile suspended solids (VSS) of *Gordonia* spp. (but not "*M. parvicella*") increased when foam appeared, and decreased when foam disappeared. This study also suggested that filamentous microorganisms need to be present above a threshold level for foam to appear in activated sludge. This possibility had been mentioned previously by [6]. Since then, two studies have determined thresholds for the occurrence of foam using in situ counting methods: Cha et al. [5] found a threshold of 1×10^6 intersections g^{-1} VSS using a filament intersection counting method and Davenport et al. [13] identified a threshold of 2×10^6 mycolata cells ml^{-1} using a quantitative FISH method. These reports did not differentiate between thresholds for foam initiation and foam stability, and the relationship of filament levels to these two phenomena remains unclear.

The present study was conducted to investigate the relationship between levels of foam-causing organisms and foam initiation and stability in activated sludge. Batch tests, lab scale reactor experiments, and analyses of a full-scale activated sludge system were used to identify and verify foaming thresholds. SSU rRNA-targeted hybridization probes were used to quantify the levels of mycolata, *Gordonia* spp. and *G. amarae*, thus avoiding the limitations of traditional identification methods.

2. Materials and methods

2.1. Culture conditions

G. amarae strains RBI and NM23 were kindly provided by M. Hernandez, University of Colorado. *G. amarae* RBI was batch-grown in TGY medium (10 g tryptone, 1 g glucose, 7 g NaCl, and 2.5 g yeast extract per liter; pH 7.0). *G. amarae* NM23 was batch-grown in mineral salts medium [14] with 0.5% acetate as the sole carbon source. After 14 days of growth, the cells were harvested by centrifugation for 10 min at $6800\times g$ and spent medium (i.e., used, cell-free medium) was collected for use in batch experiments. *G. amarae* NM23 cells were resuspended in 1 l spent medium before addition to the reactors (see below).

2.2. Activated sludge samples

For batch tests, approximately 30 l of activated sludge were obtained from the stabilization basin of a contact-stabilization activated sludge system at the Urbana-Champaign Sanitary District Northeast Wastewater Treatment Plant (UCSD-NE WWTP), Urbana, IL. This sludge sample was collected in February 1999, when the plant was not experiencing foaming problems.

To follow population shifts before and during a foaming episode, 50-ml grab samples of activated sludge mixed liquor and foam were obtained from the contact and stabilization basins of the UCSD-NE WWTP. Since this plant usually experiences filamentous foaming during the summer season (June–September) [12], samples were collected every 2–3 days from March to September, 1999. Total suspended solids and VSS of the activated sludge were determined using standard methods [15].

For membrane hybridizations, samples were collected and immediately transported to the laboratory for processing and storage. Samples (14 ml) were centrifuged at $2040\times g$ and the cell pellets were stored at $-80^{\circ}C$. Nucleic acids were extracted from cell pellets using a low-pH hot-phenol extraction procedure [16]. For FISH, 3 ml samples were immediately fixed with 9 ml 4% (w : v) paraformaldehyde (PFA) as described below.

2.3. Batch experiments

Activated sludge was placed in a 30-l container, aerated using ceramic diffusers, and mixed. Any foam that was formed was removed from the surface. Well-mixed activated sludge samples, 200 or 250 ml, were obtained from the bottom of the container and placed in 500-ml graduated cylinders. To determine the effect of spent and fresh growth media on foaming potential, batch tests were conducted in triplicate with the following samples (each totaling 250 ml): 250 ml of activated sludge (AS), 200 ml of AS + 50 ml fresh medium, 200 ml of AS + 40 ml fresh medium + 10 ml spent medium, 200 ml AS + 30 ml fresh medium + 20 ml spent medium, 200 ml AS + 20 ml fresh medium + 30 ml spent medium, 200 ml AS + 10 ml fresh medium + 40 ml spent medium, and 200 ml AS + 50 ml spent medium. To determine the effect of cell mass on foaming potential, batch tests were conducted in triplicate by suspending varying amounts of *G. amarae* RBI cell mass (7.4–298 mg of cells) in 50 ml of spent medium, which was subsequently mixed with 200 ml of activated sludge. Mixing was performed manually by slow inversion (three times) of the graduated cylinders to avoid air entrapment and the foaming potential tests were performed immediately after mixing. The added *G. amarae* biomass corresponded to 1.5%, 3.0%, 5.8%, 8.4%, 13.3%, 15.5%, 19.7%, 23.4%, and 38.0% of the VSS of the total mixture. Cell concentrations higher than the culture concentrations were obtained by centrifugation for 30 min at $10{,}600 \times g$.

Statistical analysis was performed using StatView 5.0 (SAS Institute, Cary, NC). The analysis of variance using Scheffe multiple pairwise comparison and Dunnett pairwise comparison with the control (no *G. amarae* cells added) was performed.

2.4. Foaming potential

Foaming potential tests were conducted by adding two tablets of Alka-Seltzer™ (two tablets contain 650 mg sodium acetylsalicylate, 3832 mg heat-treated sodium bicarbonate, and 2000 mg citric acid; Bayer Corp., Elkhart, IN) to 250 ml of well-mixed samples in 500-ml graduated cylinders [12]. To determine the foaming potential, the volume of foam generated was measured immediately. Foam stability was measured by noting the foam half-life, defined as the time it takes for half the volume of foam generated in the foaming potential test to dissipate.

2.5. Quantitative membrane hybridization

Membrane hybridizations were performed as previously described [8,17]. Membranes were hybridized with universal and specific oligonucleotide ^{32}P labeled probes (Table 1), and the resulting hybridization responses, measured using an Instant Imager (Packard Instruments, Meriden, CT), were used to determine the relative concentration of target SSU rRNA in the samples [8,17].

2.6. Determination of Gordonia mass using fluorescence in situ hybridization

From each mixture of activated sludge, fresh medium, and spent medium, 0.5 ml of sample was collected before the foaming potential test and placed in 1.7-ml centrifuge tubes. Each sample was fixed with 1 ml of 4% (w:v) PFA for 1 min at room temperature [8], and stored in phosphate buffered saline (PBS)/ethanol (1:1, v:v) until used for FISH. An oligonucleotide probe targeting the SSU rRNA of *Gordonia* spp. S-G-Gor-0596-a-A-22 [8], labeled with tetramethylrhodamine isothiocyanate (TRITC) or Cy3 was obtained from Genosys Biotechnologies, Inc. (The Woodlands, TX). FISH was performed as previously described [11], with the following modifications. Mutanolysin (Sigma, St. Louis, MO, Cat. No. M9909) at a concentration of $5000\,U\,ml^{-1}$ in 0.1-M phosphate buffer (pH 6.8) was used to make the cell wall of gram-positive bacteria more permeable to the probes [10] after serial dehydration in 50%, 80%, and 100% ethanol. The slides were incubated for 20 min at 37°C in a 50-ml centrifuge tube to avoid drying, the ethanol dehydration was repeated, and the slides were air dried. After FISH, the samples were stained with 4′-6-diamidino-2-phenylindole dihydrochloride (DAPI) [8]. Cells were visualized using a Zeiss Axioskop and appropriate dichromatic filter sets. Digital images were captured using a cooled charge coupled device (CCD) camera (Photometrics, Tucson, AZ) with a KAF 1400 CCD [8]. Filament length was measured using IPLab Spectrum v. 3.0 (Signal Analytics, Vienna, VA) using the Measure_Lengths command, with manual thresholding [11].

To determine the abundance of *Gordonia* spp. in the biomass, 30 random fields were captured (ten images each from three slide wells), and the total filament length from each image field was determined [11]. The mass of *Gordonia* spp. was determined using the following

Table 1
Probe names, probe sequences, target groups, and final wash temperature after membrane hybridization (T_w) for oligonucleotide probes used in this study

Probe name[a]	Probe sequence (5′-3′)	Specificity	T_w (°C)	Reference
S-*-Univ-1390-a-A-18	GACGGGCGGTGTGTACAA	Almost all organisms	44	[18]
S-D-Bact-0338-a-A-18	GCTGCCTCCCGTAGGAGT	Domain Bacteria	56	[19]
S-*-Myb-0736-b-A-22	CAGCGTCAGTTACTxCCCAGAG[b]	Mycolata	51	[8]
S-G-Gor-0596-a-A-22	TGCAGAATTTCACAGACGACGC	*Gordonia*	54	[8]
S-S-G.am-0205-a-A-19	CATCCCTGACCGCAAAAGC	*G. amarae*	53	[9]

[a] Probe names have been standardized according to the oligonucleotide probe database [20].
[b] x = 5-nitroindole.

equation [11]:

$$\textit{Gordonia}\text{ spp. conc. (mg VSS ml}^{-1}) = Rc \times [(L_f \times A_w)/(V \times a)], \quad (1)$$

where $Rc = 4.14 \times 10^{-10}$ mg *Gordonia* spp. VSS μm^{-1}, L_f the average total filament length per field for the 30 random images (μm), A_w the area of microscope slide sample well (μm^2), V the volume of sample applied to the microscope sample well (ml) and A the area of microscope field (μm^2).

The averages of the total filament length per field for the ten random image fields from each well were determined. The standard deviation of three averages (one average for each slide well) was calculated.

2.7. Lab-scale reactor studies

Three sequencing batch reactors (SBRs), each with an active working volume of 5 l were operated using peristaltic pumps and timers with a 3-h cycle: 90 min fill/react, 30 min react, 30 min settle, and 30 min decant phases.

Mixed liquor and return activated sludge from the UCSD-NE WWTP were combined to obtain mixed liquor suspended solids (MLSS) and mixed liquor volatile suspended solids (MLVSS) concentrations of 1705 and 1356 mg l^{-1}, respectively, added to a 30-l container, aerated using ceramic diffusers, and mixed with a laboratory mixer. Any foam that was formed was removed from the surface. Three 4-l activated sludge samples obtained from the bottom of the container, were each mixed with 1 l of spent medium (with or without *G. amarae* NM23 cells), and were added to the SBRs to obtain a final volume of 5 l. Reactor A was seeded with 5820 mg of *G. amarae* NM23 cells (equivalent to 51.8% of the VSS), Reactor B was seeded with 990 mg of *G. amarae* NM23 cells (equivalent to 15.4% of the VSS), and no *G. amarae* NM23 cells were added to Reactor C. A solids retention time (SRT) of 10 days was maintained by wasting daily from the mixed liquor. The hydraulic retention time was set at 6 h. Primary effluent from the UCSD-NE WWTP (average total BOD-$\approx$100 mg l^{-1}) was collected daily and used as influent for the reactors. The reactors were sampled for foaming potential, membrane hybridizations, and FISH every 2–4 days.

3. Results and discussion

3.1. Determination of Gordonia mass using fluorescence in situ hybridization

The mass of *Gordonia* spp. in activated sludge samples was determined using a previously developed FISH quantification procedure [11]. In this previous study, the relationship between total filament length and VSS for a pure culture of *G. amarae* was determined using image analysis after FISH had been performed. Assuming that the relationship between filament length and VSS in pure culture is the same as the one in activated sludge, the contribution of *Gordonia* spp. to the total VSS can be determined in activated sludge after filament length quantification by FISH and image analysis. Using this method, it was determined that the *Gordonia* level in a sample from non-foaming activated sludge obtained from the stabilization tank of the UCSD-NE WWTP was $(2.94 \pm 0.32) \times 10^7$ μm ml^{-1} or (0.5 ± 0.11)% of the VSS (mean $\pm$ standard error, $n = 3$ independent hybridizations).

To assess the effectiveness of the quantitative FISH method in activated sludge, known amounts of *G. amarae* RBI cells were added to this sample. The results are shown in Fig. 1. A line representing 100% recovery (slope of 1 : 1, intercept of 0.5% VSS, representing the *Gordonia* cell biomass in the initial sample) is shown for comparison. Within the 95% confidence limits of the method, added cells were recovered over a broad range of VSS levels, although there was a slight underestimation at the higher levels (>23%). For samples with high levels of filamentous biomass, it was difficult to measure filament length accurately because filaments disappeared from the microscope focal plane. As a consequence, filament length and biomass were underestimated. This underestimation may not be a problem

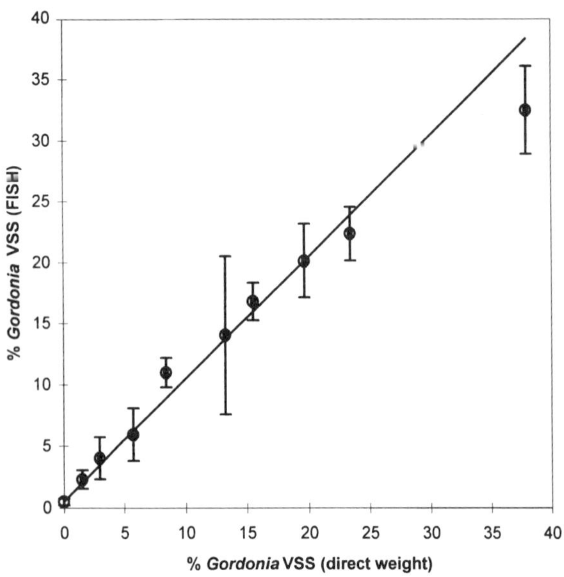

Fig. 1. Recovery of *G. amarae* RBI cells added to activated sludge samples with quantitative FISH method. The bars represent 95% confidence intervals; the line indicates the anticipated recovery (100%).

for this study since foaming activated sludge typically contained *Gordonia* levels below 20% of the VSS (see below). Thus, this quantitative FISH method can be used to quantify *Gordonia* biomass levels in activated sludge.

It should be noted that expressing *Gordonia* biomass as a percentage of total VSS may misrepresent their absolute levels in a given sample. Even if activated sludge contains a high absolute level of *Gordonia* biomass, the *Gordonia* biomass expressed as a percentage of the total VSS may be low if the total VSS is high. Expressing *Gordonia* abundance as filament length per volume or VSS concentration (mg *Gordonia* VSS l^{-1}) may be more useful for determining threshold levels, since foaming is thought to depend on the density of mycolata cells [13]. On the other hand, expressing *Gordonia* biomass as a percentage of total VSS allows us to compare the contribution of *Gordonia* biomass to total biomass with the contribution of *Gordonia* rRNA to total rRNA, thus allowing a comparison of membrane hybridization and FISH results.

3.2. Batch experiments

The foaming potential data obtained in the batch experiments are shown in Fig. 2. The foaming potential was determined to be (222 ± 1.7) ml (mean ± standard error, $n = 3$ foaming potential tests) for the undiluted activated sludge sample (3167 mg l^{-1} MLSS) and 217 ± 4.4 ml for a diluted (with distilled water) activated sludge sample (2534 mg l^{-1} MLSS). The foaming potential increased to 275 ± 2.9 ml for the sample diluted with 50 ml of fresh TGY medium. This result shows that the medium used to culture *G. amarae* RBI contributed significantly to the foaming potential, likely due to the presence of proteins or other surface active compounds in tryptone and yeast extract. A foaming potential of 250 ml had previously been shown to be the cutoff level for foaming in the UCSD-NE WWTP [12]. Thus, the addition of fresh medium during the batch tests increased the foaming potential to levels above this cutoff. To determine the effect of spent medium, varying amounts of spent medium were exchanged for fresh medium. Increasing the volume of spent medium did not change the foaming potential significantly compared to the foaming potential for the activated sludge sample diluted with fresh medium (ANOVA, $\alpha = 0.01$) (Fig. 2), indicating that the properties of the medium contributing to foaming potential did not change during cell growth. To minimize the effect of the medium on foaming potential, all cell mass additions were resuspended in 50 ml of spent medium.

Additions of pure cultures of *G. amarae* RBI of 1.5% and 3.0% of the VSS (determined by direct VSS measurement, equivalent to 1.09×10^8 and $1.94 \times 10^8\,\mu m\,ml^{-1}$, respectively) did not increase the foaming potential significantly (Fig. 2). However, at a *G. amarae* biomass level of $2.96 \times 10^8\,\mu m\,ml^{-1}$ (representing a *G. amarae* biomass concentration of 5.8% of the VSS), the foaming potential increased significantly (ANOVA, $P < 0.01$) to 333 ± 7.3 ml (mean ± standard error, $n = 3$ foaming potential tests). Thus, it is likely that a threshold level for *foam formation* corresponds to a *G. amarae* filament length between 1.94×10^8 and $2.96 \times 10^8\,\mu m\,ml^{-1}$ (3% and 5.8% of the VSS, respectively). This *G. amarae* biomass level is consistent with a 1997 study of the UCSD-NE WWTP ([12]; unpublished data), which showed an increase in *Gordonia* from less than $1.01 \times 10^8\,\mu m\,ml^{-1}$ (4% of the VSS) in non-foaming sludge to $8.52 \times 10^8\,\mu m\,ml^{-1}$ (almost 10% of the VSS) after foam initiation.

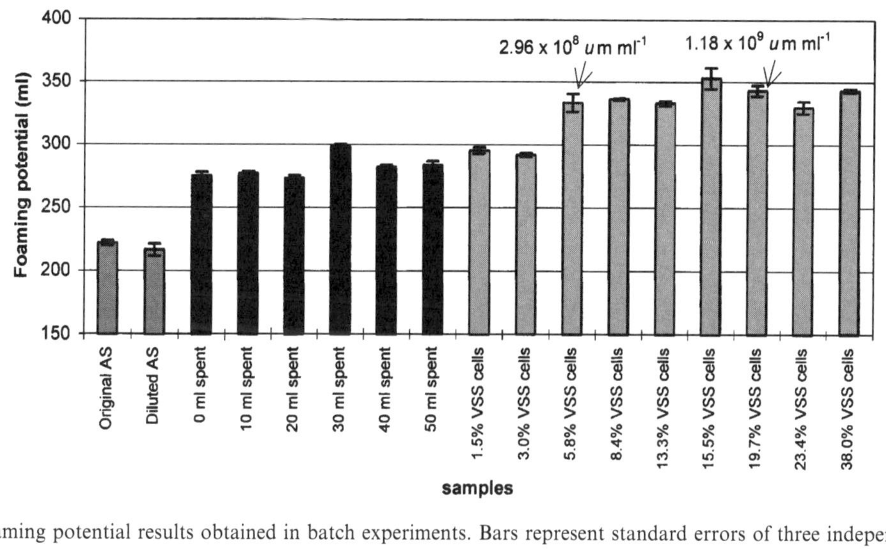

Fig. 2. Foaming potential results obtained in batch experiments. Bars represent standard errors of three independent tests.

Increasing the *G. amarae* biomass above 5.8% of the VSS did not result in significant changes (ANOVA, $P<0.01$) in the foaming potential (the foaming potential varied from 333 ± 18.7 to 353 ± 21.4 ml) (Fig. 2). However, the foam half-life dramatically increased with further increases in the *G. amarae* biomass levels (data not shown). The foam half-life increased from 60 to 90 s to 510 to 540 s when *G. amarae* was increased from 9.35×10^8 to 1.18×10^9 μm ml^{-1} (15.5% to 19.7% of the VSS). Thus, a filament level between these two values may correspond to a threshold for *foam stability*.

The first threshold is a formation threshold, which is characterized by a sudden increase in foaming potential. The second threshold is a stability threshold and we hypothesize that this threshold corresponds to a situation in which foam is maintained as a thick viscous layer on the surface of the aeration basin.

3.3. Foaming potential and stability in a full-scale activated sludge plant

To evaluate whether the formation and stability thresholds determined by batch experiments are also relevant for full-scale activated sludge systems, the UCSD-NE WWTP was studied. This plant treats primarily domestic wastewater with an average flow of 5.7×10^4 m^3 day^{-1} (15 million gallons per day) and an average BOD of 150 mg l^{-1}. The main secondary treatment process is a contact-stabilization activated sludge system, which is schematically represented in Fig. 3. Each summer, the system is temporarily switched to conventional activated sludge mode to facilitate routine annual maintenance. This plant experiences seasonal foaming, generally from June to September, with a maximum foam thickness of 6–10 cm and a maximum foam coverage of 80–100% of the aeration basin surface. Initial applications of quantitative membrane hybridizations, FISH, and immunostaining showed an approximately five-fold increase in *Gordonia* spp., rRNA, and biomass levels during a change from non-foaming to foaming conditions [12]. To describe the nature of this increase more accurately, the activated sludge plant was monitored every 2–3 days over a period of 6 months during the foaming season.

Fig. 4a shows the MLSS levels in the contact and stabilization basins and the temperature in the aeration basin immediately before the secondary clarifier (Basin D). The plant was operated in contact-stabilization mode until May 14, when the operation was switched to conventional activated sludge mode. The sampling regimen (sampling from Basins A and B) was continued after the switch in operation (MLSS levels were only determined for Basin B). Before this switch, a drastic decrease in stabilization basin MLSS occurred due to accidental prolonged wasting from the secondary clarifier. An old wasting valve was inadvertently left

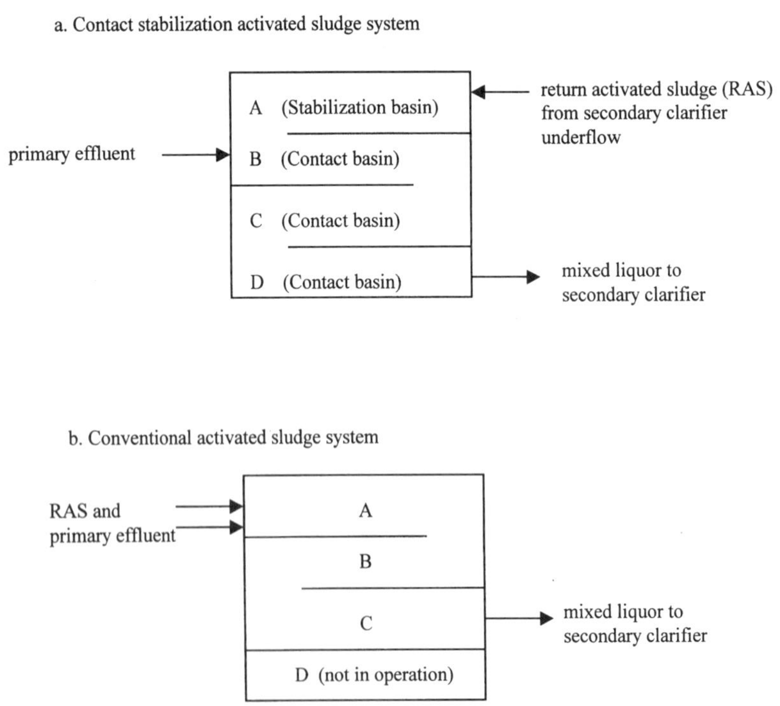

Fig. 3. Schematic representation of UCSD-NE WWTP activated sludge system. (a) Contact-stabilization activated sludge mode. (b) Conventional activated sludge mode.

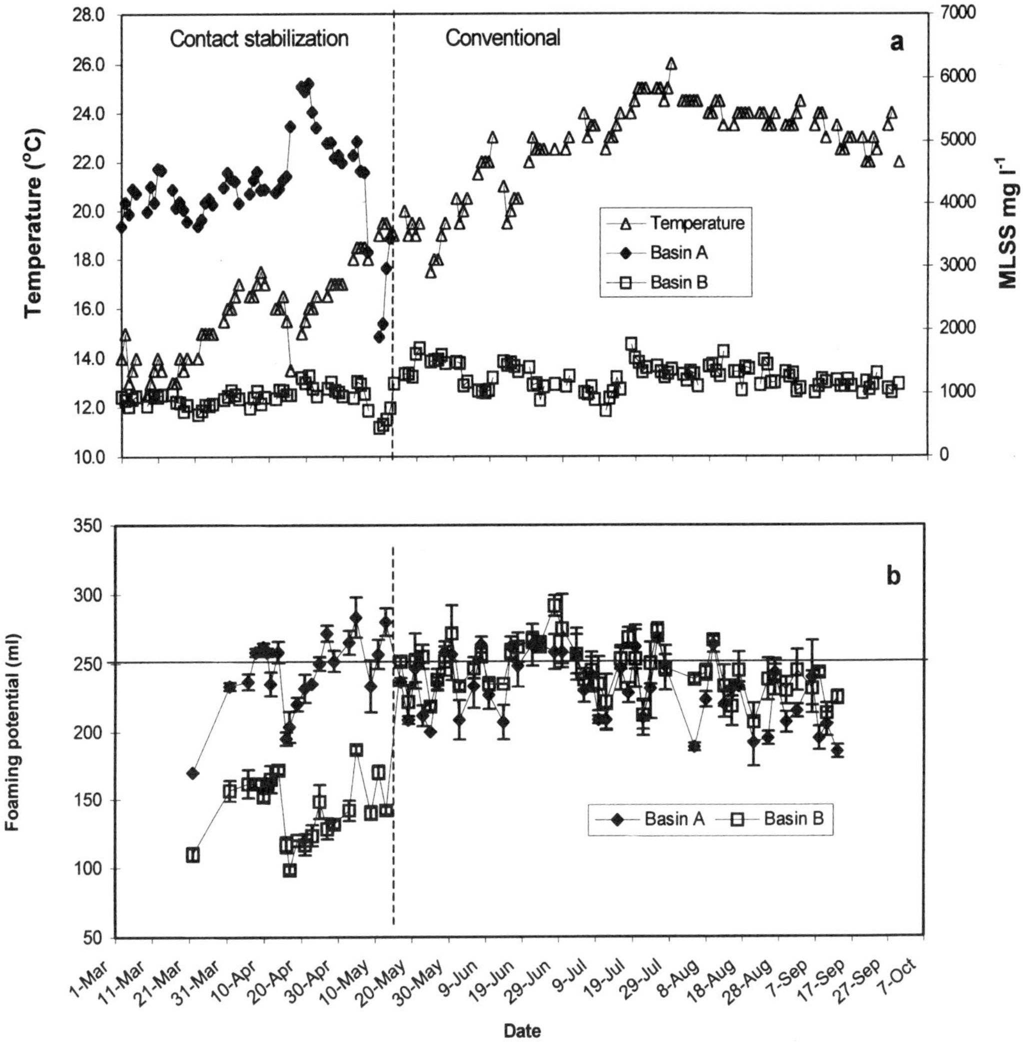

Fig. 4. Operational data for Basins A and B of the UCSD-NE WWTP activated sludge system (March–October 1999). (a) Temperature and MLSS concentrations. (b) Foaming potential

open from May 6 to 10, resulting in a massive loss of solids from the system, particularly from Basin A (Fig. 4a).

Fig. 4b shows the foaming potential over time. When sampling began on March 22, the foaming potential in Basin A was initially low (less than 200 ml), which is consistent with the absence of surface foam at the plant. The foaming potential then increased and reached 250 ml on April 8, the cutoff value for foam formation for this plant [12]. At the same time, some foam (less than 5% surface coverage of Basin A) was observed. The increasing values of the foaming potential immediately before the appearance of surface foam indicate that monitoring foaming potential can be part of an early warning system for predicting the onset of foaming. The surface foam rapidly disappeared following two rainfall events (April 8 and 10), resulting in a drop in foaming potential. This observation indicates that the foam was unstable and that the hydraulic pattern in the basins was only partially foam-trapping. Subsequently, the foaming potential increased to a value greater than 250 ml on April 25. The foaming potentials of the contact basin and the stabilization basin during this period behaved similarly, although the foaming potential in the contact basin was lower because of the lower solids concentration. Surface foam was again observed on April 25 and remained present during the rest of the sampling period, while surface foam coverage varied from 10% to 95%. The foaming potential for the rest of the sampling period fluctuated from 185 to 292 ml. As expected, the foaming potentials for the two basins became very similar after the switch to conventional activated sludge mode. From June 13 to June 27, the plant operators tried to control the foam by manipulating the SRT. The waste activated sludge (WAS) flow rates were increased to lower the SRT on June 14, decreased on June 24 as the MLSS dropped below $1000\,\mathrm{mg\,l^{-1}}$, and increased again on June 27. This resulted in a decrease in foaming potential after June 29, but surface foam was not completely eliminated, and the foaming potential returned to its previous levels by mid-July.

During the entire monitoring period, the foam half-life never exceeded 180 s, and was mostly around 10–20 s. This is well below the half-life of 510–540 s, corresponding to the stability threshold determined in batch tests. Therefore, the foam-causing filaments were probably present at a concentration below this stability threshold, and the filament levels in this activated sludge plant were likely between formation and stability thresholds during foaming episodes.

3.4. Changes in mycolata rRNA levels during foaming in a full-scale activated sludge plant

The results of membrane hybridizations with the mycolata-, *Gordonia*-, and *G. amarae*-specific probes are shown in Fig. 5. In general, the levels were relatively low in the mixed liquor of both aeration basins (less than 5% of the total SSU rRNA). During the first period of observation (March 22–May 11), the mycolata rRNA percentages were considerably higher than the *Gordonia* and *G. amarae* rRNA percentages; *Gordonia* levels were approximately 0.5% and *G. amarae* levels were close to the limit of detection (LOD) of 0.4% (in Fig. 6, values below the LOD are presented as one half of the LOD [i.e., 0.2%], following the suggestion of Clarke [21]). Since surface foam was first observed in early April, the initial foam was not linked to high levels of *Gordonia* rRNA, but may be due to the presence of other mycolata. Following two rainfall events on April 8 and 10, the foam disappeared, and the foaming potential decreased (Fig. 4b). This decrease in foaming potential coincided with a decrease in mycolata rRNA levels (particularly for Basin A) suggesting that (1) the initial foam was due to mycolata cells other than *Gordonia*, or mycolata-produced surfactants, and (2) this foam was not stable, and easily affected by changes in the hydraulic patterns in the plant. After this temporal washout event, the mycolata rRNA levels in both basins increased again. The mycolata rRNA levels declined once more during inadvertent wasting (May 5–13). After the switch to conventional activated sludge mode, the *Gordonia* and *G. amarae* rRNA fractions slowly increased, and this increase paralleled the rise in basin temperature from approximately 15–20°C (Fig. 4a). The mycolata, *Gordonia*, and *G. amarae* rRNA fractions continued to increase and reached values of 3.3%, 2.2%, and 2.1% of the total SSU rRNA, respectively, in Basin B, and 3.6%, 1.9%, and 1.9%, respectively, in Basin A, on June 16. Active manipulation of the WAS flow rates to reduce the foaming (see above) disturbed this increasing trend and probably resulted in the decrease of the mycolata, *Gordonia*, and *G. amarae* rRNA fractions after June 16. After this disturbance, the mycolata and *Gordonia* rRNA percentages were very similar (mycolata levels were sometimes even lower than *Gordonia* rRNA percentages, but the differences were small given the variability of the quantitative hybridization assay [17]). These results again demonstrate the relative susceptibility of non-*Gordonia* mycolata to operational changes. The increase in *Gordonia* rRNA levels shows that foaming was probably due to *Gordonia* after this point. While the *G. amarae* rRNA levels followed the same trend as those of *Gordonia*, the presence of one or more *Gordonia* spp. other than *G. amarae* is suggested and may signify the possible role of other *Gordonia* spp. in foaming. The rRNA percentages for the mycolata, *Gordonia*, and *G. amarae* decreased considerably on September 10, consistent with decreasing intracellular rRNA levels determined by FISH (see below).

The mycolata, *Gordonia*, and *G. amarae* rRNA levels in foam are shown in Fig. 5c. From March 22 to May 11, the levels of mycolata rRNA in foam were close to those in the mixed liquor. These results suggest that the initial foaming (before May 11) was not due to selective enrichment of mycolata cells in foam, but may be caused by increased biosurfactant production by mycolata. The sudden disappearance of foam in response to hydraulic events is consistent with biosurfactants being washed out from the system. After May 11, mycolata cells were selectively enriched in the foam (Fig. 5c). This observation corresponded to the increases in mycolata, *Gordonia*, and *G. amarae* rRNA levels in the mixed liquor. During this period, the mass of cells in the foam, and not simply the presence of biosurfactant, appeared to be the cause of foaming. This foaming was also more stable, i.e., not easily affected by changes in WAS flow rates or SRTs, as evidenced by (1) the maintenance of relatively high rRNA levels, and (2) the continued presence of surface foam. This verifies that once cells are enriched in the foam, the manipulation of SRT (within normal operating limits) may not be effective in completely removing the foam, because most of the cells in the foam layer are not removed by normal wasting procedures. This selective enrichment occurred when the temperature was higher than 17°C, supporting the hypothesis of higher *Gordonia* growth in activated sludge at higher temperatures, and may explain the yearly occurrence of foaming at this plant.

The operational adjustments in WAS flow beginning on June 14 resulted in a decrease in mycolata, *Gordonia*, and *G. amarae* rRNA levels in the foam, but did not result in the complete removal of surface foam. As stated above, the complete removal of mycolata in foam

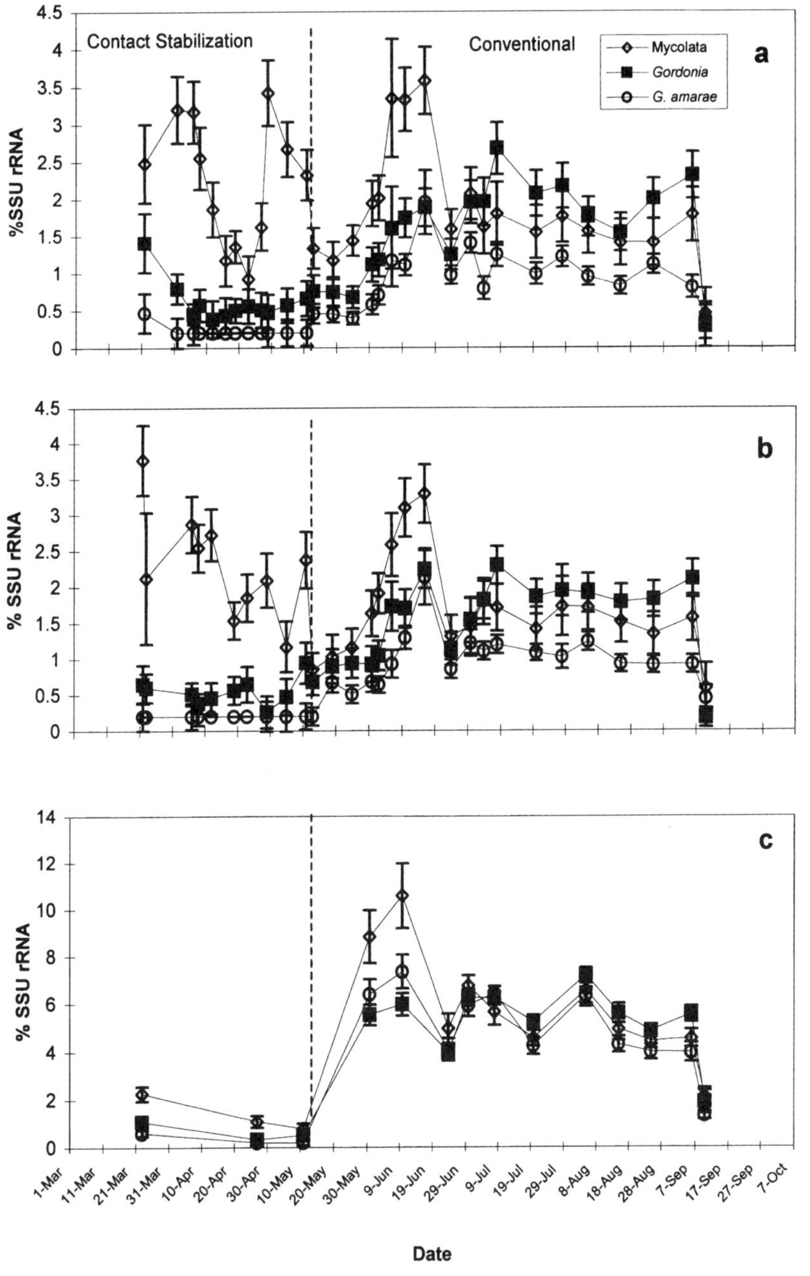

Fig. 5. Results of quantitative membrane hybridizations with oligonucleotide probes for the mycolata, *Gordonia*, and *G. amarae* for UCSD-NE WWTP. (a) Basin A mixed liquor. (b) Basin B mixed liquor. (c) Basin B foam.

by SRT manipulation may not be possible without compromising plant operation. During the rest of the summer, the rRNA levels in the foam did not fluctuate much and *Gordonia*, particularly *G. amarae*, constituted most of the mycolata in the foam. Before this period of instability, mycolata other than *Gordonia* were present in the foam.

Considering all perturbations during the entire period, it can be concluded that mycolata other than *Gordonia* were more sensitive to operational changes and that *Gordonia* spp. particularly *G. amarae*, became more dominant over the course of foaming. This may explain the historical attribution of most foaming phenomena to *G. amarae*, particularly in the United States. Not only

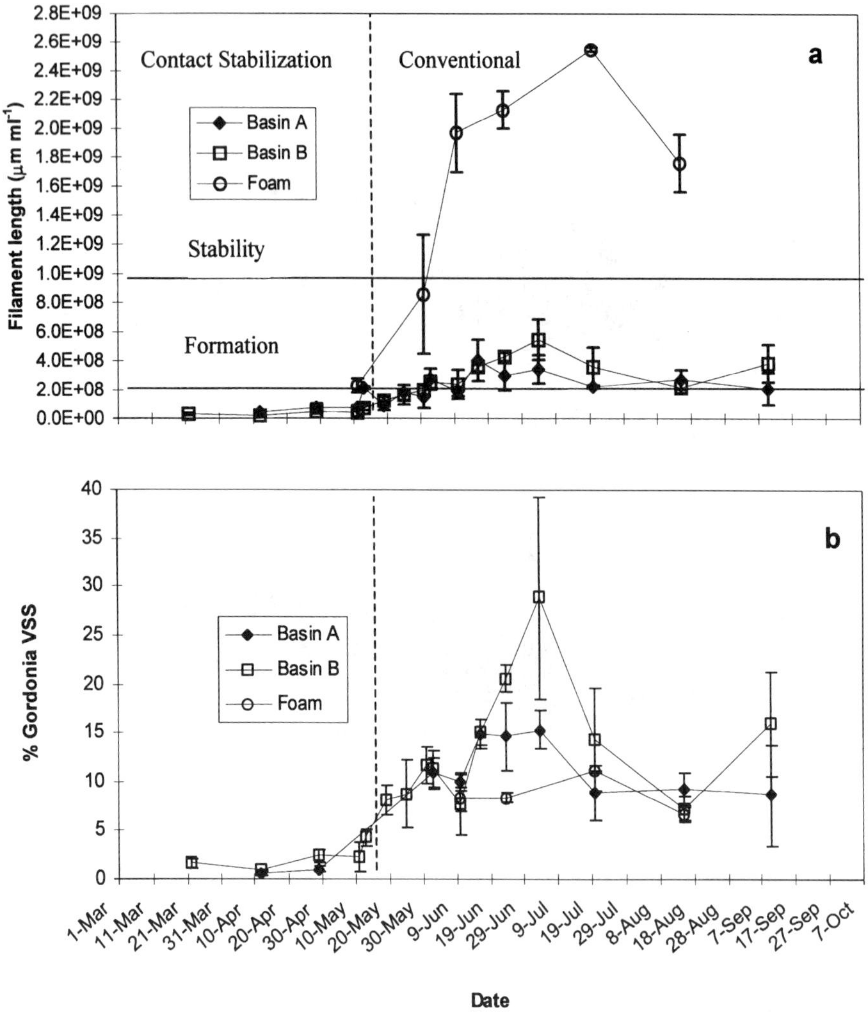

Fig. 6. Quantitative FISH results for mixed liquor and foam from the UCSD-NE WWTP. (a) *Gordonia* filament length per volume (b) % *Gordonia* VSS.

did *Gordonia* become the dominant mycolata genus as foaming progressed, it was also less susceptible than other mycolata to sudden operational perturbations such as changing flows and wastage rates.

3.5. Changes in Gordonia biomass during foaming in a full-scale activated sludge plant

The levels of *Gordonia* biomass determined by quantitative FISH are shown in Fig. 6. Fig. 6a shows the levels expressed as total *Gordonia* filament length per sample volume ($\mu m\, ml^{-1}$). In March and April, the total *Gordonia* filament lengths in basins A and B varied between 1.9×10^7 and $7.9 \times 10^7\, \mu m\, ml^{-1}$. These filament levels were below the threshold level for foam formation determined by batch tests (approximately $2 \times 10^8\, \mu m\, ml^{-1}$). The total filament length increased as foaming progressed and peaked on July 4 at $5.5 \times 10^8\, \mu m\, ml^{-1}$ (basin B). Comparable values were observed during the same time period in the same activated sludge plant in 1997, with foaming sludge having total filament lengths greater than $2 \times 10^8\, \mu m\, ml^{-1}$ (unpublished data). Taken together, these data suggest that a total filament length of $2 \times 10^8\, \mu m\, ml^{-1}$ in the mixed liquor is a possible threshold for foam formation. The data for this plant indicate that this first threshold was reached in late May (Fig. 6a). The *Gordonia* filament levels in the foam also increased considerably during late May and early June.

The second threshold determined in batch tests, the foam stability threshold (approximately $1 \times 10^9\, \mu m\, ml^{-1}$), was not reached in this plant during the 1999 foaming season (Fig. 6a). This was further demonstrated by the low foam half-life values measured for this plant (<10–180 s). This second threshold was evaluated in more detail using lab-scale reactor tests (see below).

The filament length measurements determined by quantitative FISH were converted to VSS levels of *Gordonia* according to Reyes et al. [11] and normalized using the total VSS of the activated sludge or foam samples. Using this approach, the results for the UCSD-NE WWTP can be expressed as % *Gordonia* VSS (Fig. 6b). A comparison of the values of % rRNA (Fig. 6) and % *Gordonia* VSS (Fig. 6b) shows similar four- to seven-fold increases of % *Gordonia* rRNA (<0.5–2%) and % *Gordonia* VSS (4–28%) during foaming. In the period before the start of "true" filamentous foaming, the *Gordonia* % rRNA and % VSS values were low (0.6% and 4.2%, respectively). However, at its peak value, the *Gordonia* contribution, of 28%, to the VSS in Basin B corresponded to only 2% of the total rRNA. This discrepancy may be due to: (1) overestimation of the VSS contribution of *Gordonia*; and/or (2) relatively low rRNA levels of *Gordonia* cells in mixed liquor during foaming. The overestimation of *Gordonia* biomass is possible, since the VSS conversion assumes that the VSS of *Gordonia* grown in pure culture is equivalent to the VSS of *Gordonia* growing in activated sludge [11]. This assumption may not be correct, given that the cell envelope composition and surface-active lipids of the mycolata are dependent on substrate and other cultivation conditions [22,23]. In addition, the filaments in activated sludge appeared thinner than the pure culture cells used to develop the *Gordonia* filament length to VSS conversion. Therefore, it is likely that the *Gordonia* VSS concentration per unit filament length in activated sludge is lower than that in pure culture. More detailed comparative measurements of filament width (diameter) need to be performed to address this possible limitation. The second scenario (low rRNA numbers in *Gordonia* cells) may also be possible, given previous studies that compare FISH and immunostaining results. The simultaneous application of FISH and immunostaining in activated sludge showed variable and weak FISH fluorescent signals within the same *Gordonia* filament [24], indicating low target rRNA levels in *Gordonia* cells. In addition, filament lengths obtained using fluorescently labeled antibody stains are generally higher than FISH estimates in various activated sludge plants [11].

It remains to be resolved whether the low rRNA levels in *Gordonia* cells are intrinsic, or due to low growth activity. Recently, it has been shown that there may be as much as a 15-fold difference in SSU rRNA content per cell in different species of marine Proteobacteria growing at similar growth stages [25]. It is possible that *Gordonia* have intrinsically low cellular rRNA concentrations compared to other activated sludge organisms.

It is also important to note that the filament levels (expressed as $\mu m\,ml^{-1}$) in the foam are five to ten times higher than those in the mixed liquor (Fig. 6a), while the rRNA percentages in foam and mixed liquor show a three- to five-fold difference (Fig. 5). However, using the quantitative FISH method to estimate *Gordonia* biomass shows comparable rRNA and VSS percentages in the foam. In this case, normalization to the foam VSS may not be an accurate method for expressing the *Gordonia* contribution to biomass, as the foam layer contains much more non-biomass solids than the mixed liquor. Again, the filament length per unit volume may be a more meaningful representation of *Gordonia* biomass levels than % VSS.

The FISH analysis visually revealed the changes in cell morphologies as foaming progressed. In March and April, *Gordonia* cells were mostly present as isolated short rods inside activated sludge flocs. In mid- to late-May, large clusters of rod-shaped *Gordonia* cells were observed inside flocs (Figs. 7a and b). These cell clusters were observed in increasing frequency towards the end of May and early June, and the cells were brighter than the typical *Gordonia* cells observed in March and April. By early June, typical branching filaments were observed in the mixed liquor and the foam (Fig. 7c). These filaments were much brighter than the filaments observed in July and August. Quantification of hybridization signals in individual cells is beyond the scope of this study. However, it is interesting to note that the brightest signals were observed in late May and early June, when *Gordonia* growth rates were presumably highest. During the end of the sampling period (September 10), low signal intensities within filaments were observed (Fig. 7d). This observation is consistent with the decline in mycolata, *Gordonia*, and *G. amarae* rRNA percentages as measured by membrane hybridization (Fig. 5). However, the reason for this decrease in rRNA abundance is unclear.

3.6. Lab-scale reactor studies

Using batch experiments, we identified a stability threshold (see above). However, we did not observe an increase in foam stability in the full-scale plant during the entire study period, and hypothesized that the stability threshold was never reached. During previous, unsuccessful attempts to induce filamentous foaming in lab-scale reactors under similar conditions (unpublished data), we had learned that it was difficult to maintain *Gordonia* in lab-scale systems for extended periods of time. Therefore, we decided to verify the stability threshold value as the foam disappeared (i.e., washout of *Gordonia* cells). The levels of *Gordonia* cells added to the reactors were chosen to simulate different conditions with respect to the thresholds identified by the batch tests: Reactor A was seeded with *Gordonia* cells to operate above the stability threshold (51.8% of the VSS), Reactor B was run below the stability but above

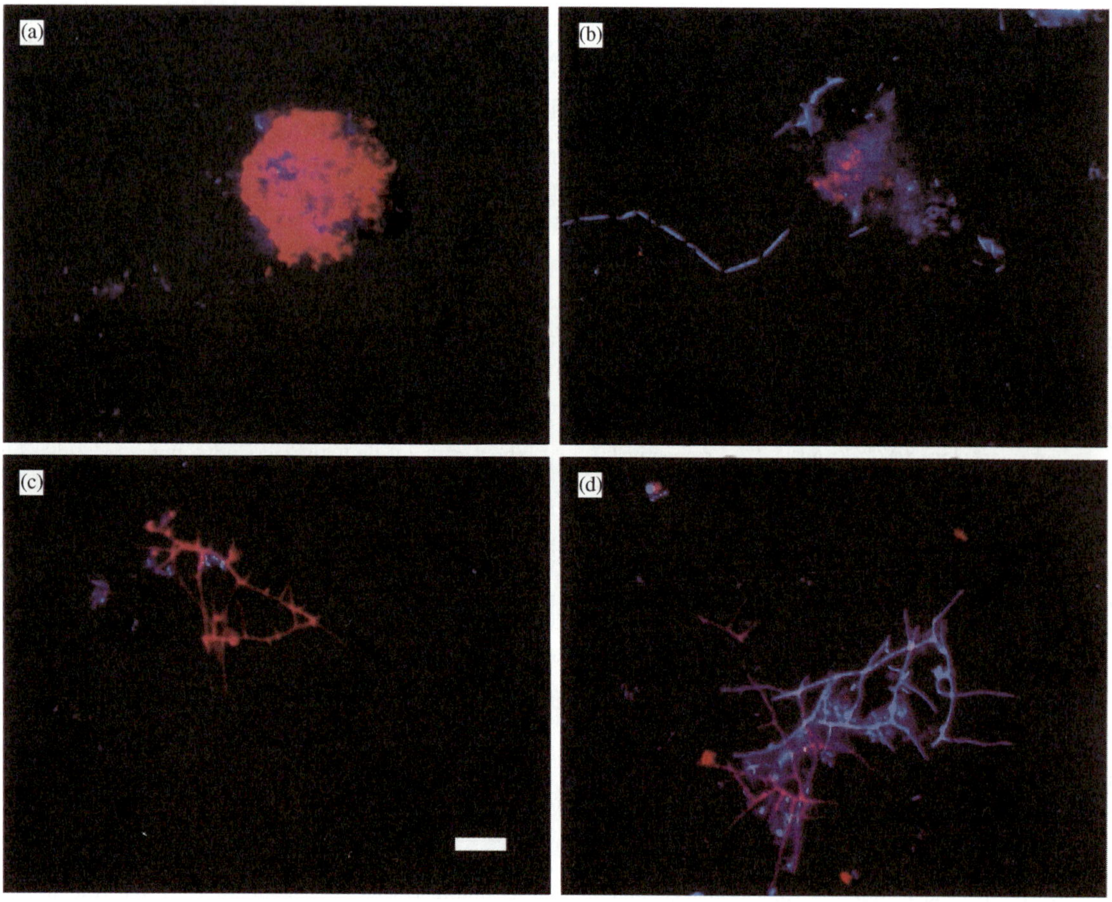

Fig. 7. Epifluorescence micrographs of *Gordonia* in activated sludge foam from the UCSD-NE WWTP. Samples were hybridized with a TRITC-labeled probe S-G-Gor-0596-a-A-18 (shown in red) and were simultaneously stained with DAPI (shown in blue). (a) and (b) May 11, 1999. (c) May 31, 1999. (d) September 10, 1999. The length of the bar corresponds to 10 μm.

the foam formation threshold (15.4% of the VSS), and Reactor C was operated below the foam formation threshold (no cells added; quantitative FISH showed that the *Gordonia* filament levels were $1.7\times10^{8}\,\mu m\,ml^{-1}$). The reactors were operated until surface foam disappeared from Reactor A (14 days).

Immediately after the *Gordonia* cells were added, Reactors A and B developed foam. The foam in Reactor A increased to a maximum height of 36 cm (foam volume of 5.5 l), while Reactor B foam rose to a maximum height of 6.5 cm (foam volume of 1 l). To ensure *Gordonia* washout (i.e., to prevent foam trapping), the foam was reincorporated into the mixed liquor several times a day by manual mixing. The foaming potential and corresponding foam half-life values are shown in Fig. 8a. Reactor C exhibited minimal surface foam, as expected, since the filament level was below the hypothesized foam formation threshold of approximately $2\times10^{8}\,\mu m\,ml^{-1}$. To obtain conservative values, Reactor C's initial foaming potential of 207 ml was taken as the lower value for foam formation. From this analysis, Reactor B crossed this foam formation threshold between Days 6 and 9 and Reactor A crossed this threshold between Days 11 and 14. The foam half-life data for Reactor A decreased between Days 3 and 6 and crossed the foam stability threshold during this time period (Fig. 8a). Reactors B and C did not exhibit any significant foam stability, which was consistent with the levels of cells added to the reactors.

The *Gordonia* biomass levels in the mixed liquor (expressed as filament length per sample volume) are shown in Fig. 8b. On Days 6 and 9, filament lengths of 3.5×10^{8} and $5.92\times10^{7}\,\mu m\,ml^{-1}$, respectively, were observed in Reactor B. On Days 11 and 14, filament lengths of 1.95×10^{8} and $1.83\times10^{7}\,\mu m\,ml^{-1}$, respectively, were observed in Reactor A. Therefore, the foam formation threshold appears to be slightly below $2\times10^{8}\,\mu m\,ml^{-1}$. Again, this is consistent with the batch tests (foam formation threshold of

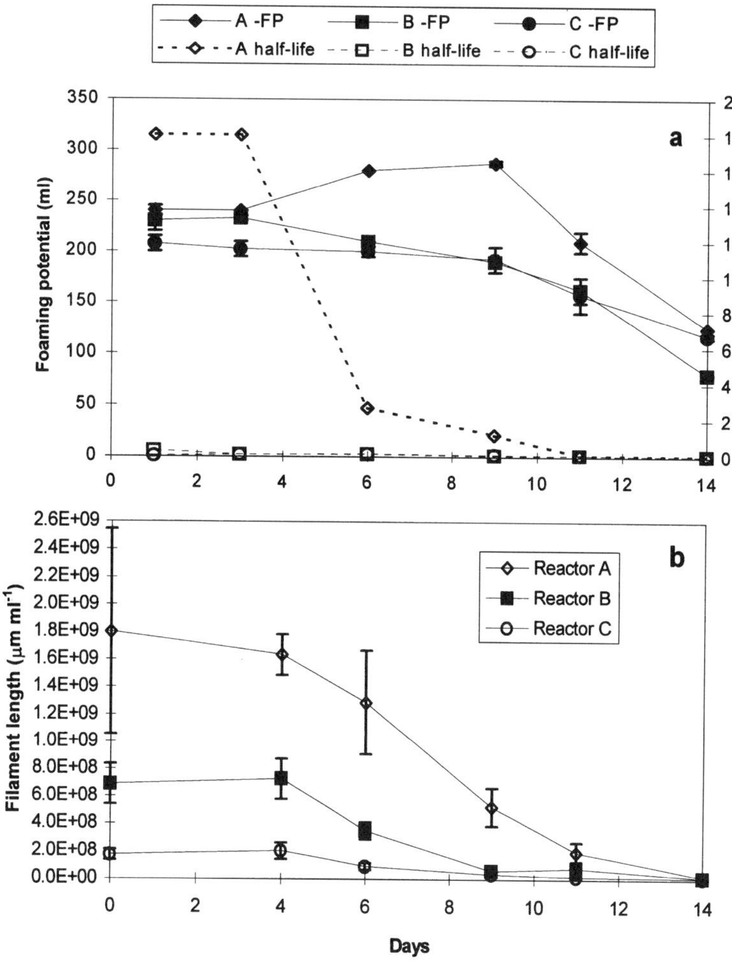

Fig. 8. Lab-scale reactor data. (a) Foaming potential and stability. (b) *Gordonia* biomass determined by quantitative FISH.

$>1.94\times10^{8}\,\mu m\,ml^{-1}$), 1997 data from the UCSD-NE WWTP ([12]; unpublished data), and 1999 data from the UCSD-NE WWTP presented in this study. On Day 6, the total filament length for Reactor A was $1.29\times10^{9}\pm0.16\times10^{9}\,\mu m\,ml^{-1}$. The value of $1\times10^{9}\,\mu m\,ml^{-1}$ can be taken to be the lower bound of the foam stability threshold. We hypothesize that, because the filament level in the activated sludge of the UCSD-NE WWTP did not exceed this second threshold, foaming at this plant is not as stable as in other plants where foam is present throughout the year.

The rRNA fractions of the mycolata are shown in Fig. 9. The washout of *G. amarae* from all three reactors was observed. In addition, the data indicate that the contribution of *G. amarae* to the total rRNA was low. For example, approximately 10–12% of the total rRNA was found to be *G. amarae* on Day 0 in Reactor A, compared to the addition of *G. amarae* biomass of 51.6% of the total VSS. This again suggests that the intracellular rRNA concentrations in *G. amarae* are low compared to those in other activated sludge organisms.

4. Implications and conclusions

Since the crucial aspect in understanding foaming is foam initiation, it is necessary to quantify the levels of causative organisms before and during foaming. In this study, the relationship of *Gordonia* levels to foaming was evaluated using oligonucleotide hybridization probes that allow identification and quantification without the limitations of traditional microscopy techniques. Batch tests provided evidence for the cause-effect relationship of *G. amarae* to foaming. These tests also identified two possible thresholds: a formation threshold and a stability threshold. While the existence of two thresholds may be a universal phenomenon in foaming, it is possible that the filament levels associated with these

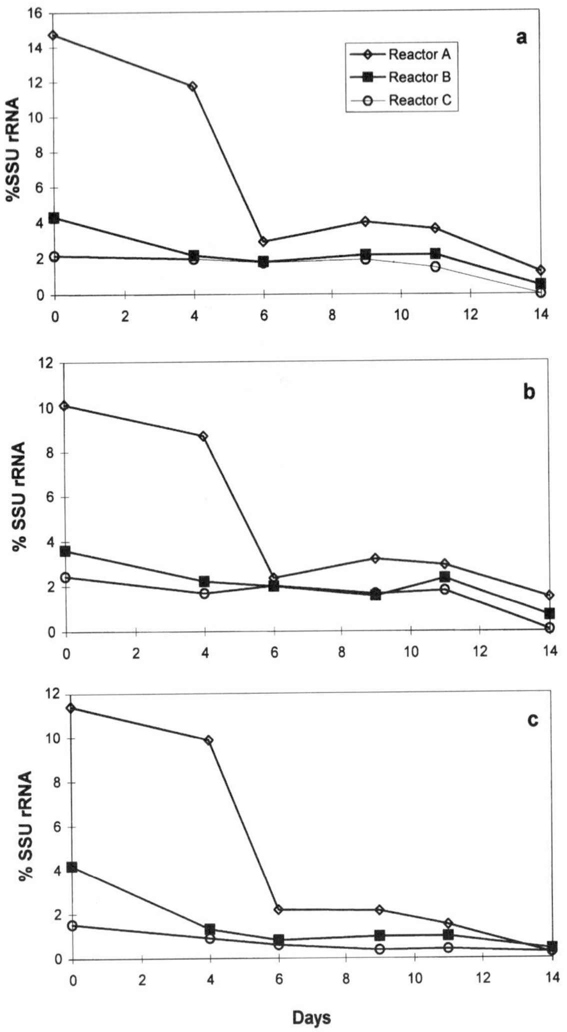

Fig. 9. Results of quantitative membrane hybridizations with oligonucleotide probes for the mycolata (a), *Gordonia* (b), and *G. amarae* (c) for the lab-scale reactors.

thresholds are treatment plant-specific. Once these threshold levels are determined for a given treatment plant, then foam occurrence and stability can be predicted based on *Gordonia* levels. This may allow the objective evaluation of incremental effects of control methods. For example, a particular control method may be effective in lowering *Gordonia* levels, but the effect can be masked by the continued presence of foam. The quantification of *Gordonia* levels in relation to threshold levels may allow the identification of crucial stages (e.g., from stable to unstable foam) during the control period. However, it should be noted that other factors, such as changes in cell surface hydrophobicity or biosurfactant production, may affect the threshold values. Additional studies are needed to determine the role of these factors in foam initiation and stability.

The monitoring of a full-scale plant over the course of foaming suggested that initial foaming in this plant was induced by mycolata other than *Gordonia*, possibly by biosurfactant production and not cell mass. During this initial foam period, the levels of mycolata in the foam layer were similar to those in the mixed liquor, indicating that selective enrichment of mycolata in the foam layer did not occur. In addition, the sudden disappearance of the foam following changes in plant hydraulics is consistent with biosurfactant washout from the system.

As foaming progressed, the growth of *Gordonia*, and particularly *G. amarae*, was favored in the mixed liquor and in the foam. This was apparently related to increasing temperature and the susceptibility of non-*Gordonia* mycolata to operational perturbations, such as sudden changes in plant hydraulics and solids inventory. Once a critical value of *Gordonia* biomass was surpassed (formation threshold), "true" filamentous foaming occurred, and the manipulation of SRT was not completely effective in removing the foam.

The stability threshold is likely relevant for plants with a continuous foaming problem. Activated sludge plants that experience seasonal *Gordonia* foaming may have *Gordonia* levels between the formation and stability thresholds. Systems with levels below the formation threshold may have non-*Gordonia* foaming due to the presence of other mycolata or biosurfactants. These systems are characterized by the sudden appearance and disappearance of foam in response to system perturbations. The combined use of foaming potential tests and oligonucleotide probe hybridizations can characterize activated sludge plants in terms of these thresholds. This approach could provide an early warning system for foam occurrence or stability and result in a guide for evaluating the effectiveness of foam control measures.

Acknowledgements

We are thankful to Tim Bachman, Mike Guthrie and the operators of the Urbana-Champaign Sanitary District for their support. Valuable discussions with Glen Daigger, Mark Hernandez, Dan Oerther, Dominic Frigon, Margit Mau, and the assistance of Dominic Frigon with statistical analysis are appreciated. This research was supported by the U.S. National Science Foundation (Grants BES 9410476 and BES 97-33826).

References

[1] Goodfellow M, Chun J, Stubbs S, Toboli AS. Transfer of *Nocardia amarae* Lechevalier and Lechevalier 1974 to the genus *Gordona* as *Gordona amarae* comb nov. Lett Appl Microbiol 1994;19:401–5.

[2] Goodfellow M, Alderson G, Chun G. Rhodococcal systematics: problems and developments. Antonie van Leeuwenhoek 1998;74:3–20.
[3] Soddell JA, Seviour RJ. A review: microbiology of foaming in activated sludge plants. J Appl Bacteriol 1990;69:145–76.
[4] Jenkins D, Richard M, Daigger GT. Manual on the causes and control of activated sludge bulking and foaming. Chelsea, Michigan: Lewis Publishers, Inc, 1993.
[5] Cha DK, Jenkins D, Lewis WP, Kido WH. Process control factors influencing *Nocardia* populations in activated sludge. Water Environ Res 1992;64:37–43.
[6] Pitt P, Jenkins D. Causes and control of *Nocardia* in activated sludge. Res J Water Pollut Control Fed 1990;62:143–50.
[7] Dhaliwal BS. Discussion of: causes and control of *Nocardia* in activated sludge. Res J Water Pollut Control Fed 1991;63:184–5.
[8] de los Reyes F, Ritter W, Raskin L. Group-specific small-subunit rRNA hybridization probes to characterize filamentous foaming in activated sludge systems. Appl Environ Microbiol 1997;63:1107–17.
[9] de los Reyes MF, de los Reyes FL, Hernandez M, Raskin L. Quantification of *Gordona amarae* strains in foaming activated sludge and anaerobic digester systems using oligonucleotide hybridization probes. Appl Environ Microbiol 1998;64:2503–12.
[12] Oerther DB, de los Reyes FL, de los Reyes MF, Raskin L. Microbial populations in a full scale activated sludge treatment piant before, during, and after an incident of seasonal biological foaming. Water Res 2001;35(14):3325–36.
[11] de los Reyes FL, Oerther DB, de los Reyes MF, Hernandez M, Raskin L. Characterization of filamentous foaming in activated sludge systems using oligonucleotide hybridization probes and antibody probes. Water Sci Technol 1998;37(4–5):485–93.
[12] Oerther DB, de los Reyes FL, de los Reyes MF, Raskin L. Microbial populations in a full scale activated sludge treatment plant before, during, and after an incident of seasonal biological foaming. Water Res 2001;35(14): 3325–36.
[13] Davenport RJ, Curtis TP, Goodfellow M, Stainsby F, Bingley M. Quantitative use of fluorescent in situ hybridization to examine relationships between mycolic acid-containing actinomycetes and foaming in activated sludge plants. Appl Environ Microbiol 2000;66:1158–66.
[14] Stanier RY, Palleroni NJ, Doudoroff M. The aerobic pseudomonads. A taxonomic study. J Gen Microbiol 1966;43:159–277.
[15] Greenberg AE, Clesceri LS, Eaton AD, editors. Standard methods for the examination of water and wastewater. Washington DC: American Public Health Association (APHA), 1992.
[16] Raskin L, Zheng D, Griffin ME, Stroot PG, Misra P. Characterization of microbial communities in anaerobic bioreactors using molecular probes. Antonie van Leeuwenhoek 1995;68:297–308.
[17] Raskin L, Capman WC, Sharp R, Poulsen LK, Stahl DA. Molecular ecology of gastrointestinal ecosystems. In: Mackie RI, White BA, Isaacson RE, editors. Ecology and physiology of gastrointestinal microbes, vol. 2: gastrointestinal microbiology and host interactions. New York: Chapman and Hall, 1997. p. 243–98.
[18] Zheng D, Alm E, Stahl DA, Raskin L. Characterization of universal small-subunit rRNA hybridization probes for quantitative molecular microbial ecology studies. Appl Environ Microbiol 1996;62:4504–13.
[19] Amann R, Binder BJ, Olson RJ, Chisholm SW, Devereux R, Stahl DA. Combination of 16S rRNA-targeted oligonucleotide probes with flow cytometry for analyzing mixed microbial populations. Appl Environ Microbiol 1990;56:1919–25.
[20] Alm E, Oerther D, Larsen N, Stahl D, Raskin L. The oligonucleotide probe database. Appl Environ Microbiol 1996;62:3557–9.
[21] Clarke JU. Evaluation of censored data methods to allow statistical comparisons among very small samples with below detection limit observations. Environ Sci Technol 1998;32:177–83.
[22] Lang S, Philp JC. Surface-active lipids in rhodococci. Antonie van Leeuwenhoek 1998;74:59–70.
[23] Sutcliffe IC. Cell envelope composition and organisation in the genus *Rhodococcus*. Antonie van Leeuwenhoek 1998;74:49–58.
[24] Oerther DB, de los Reyes FL, Hernandez M, Raskin L. Simultaneous oligonucleotide probe hybridization and immunostaining for in situ detection of *Gordonia* species in activated sludge. FEMS Microbiol Ecol 1999;29: 129–36.
[25] Kerkhof L, Kemp P. Small ribosomal RNA content in marine proteobacteria during non-steady-state growth. FEMS Microbiol Ecol 1999;30:253–60.

PERGAMON

Water Research 36 (2002) 460–468

WATER RESEARCH

www.elsevier.com/locate/watres

Assessment of activated sludge viability with flow cytometry

Giuliano Ziglio[a], Gianni Andreottola[a], Silvia Barbesti[b], Giorgio Boschetti[b], Laura Bruni[b], Paola Foladori[a,*], Roberta Villa[a]

[a] *Department of Civil and Environmental Engineering, University of Trento, via Mesiano, 77, 38050 Trento, Italy*
[b] *Biological and Chemical Laboratories–S.O.I.S, Province of Trento, via Gazzoletti 33, 38100 Trento, Italy*

Abstract

The aim of the study was to evaluate the applicability of fluorescent dyes and multiparameter flow cytometry for the rapid and direct viability/activity assessment of activated sludge samples taken from wastewater treatment plants. Viability and activity of the biomass were estimated respectively through cellular membrane integrity, staining with SYBR Green I and Propidium Iodide, and through fluorogenic dyes capable of detecting enzymatic activity, as FDA and BCECF-AM. A procedure has been developed to disaggregate sludge flocs before dyes staining and cytometric analysis. The developed procedure allows a high recovery of bacteria with good accuracy and repeatability, and minimize the damage of the cells suspension obtained from the disaggregation of the flocs. These measurements were applied to estimate the two main parameters required to define the biological activated sludge process: the endogenous decay rate and the specific growth rate in exponential phase with high F/M ratio. *Oxygen utilization rate* measurements (OUR) were conducted to conventionally monitor the activity of the biomass. The preliminary data are encouraging and support the possibility to investigate bacteria dynamics on wastewater treatment plants.

Keywords: Activated sludge; Flow cytometry; Fluorescent probes; Membrane integrity; Enzymatic activity; OUR; Kinetic constants

1. Introduction

Nowadays, activated sludge plants are the most widespread solution for the treatment of both municipal and industrial wastewater. Several models were developed for the simulation of activated sludge plants under steady-state and dynamic conditions [1,2]. For the application of these models it is required to know the characterization of the influent wastewater, the kinetic constants of the biological process and the active fraction of the biomass. Among the parameters needed for the models, the growth and decay rate and the value of active biomass play a fundamental role and depend on the investigated activated sludge. The biomass of an activated sludge tank is normally expressed through the concentration of volatile suspended solids (VSS); it is quite obvious that the mere measurement of VSS is not adequate to express the active biomass. The volatile suspended solids of the mixed liquor contain, besides active microorganisms, biodegradable, endogenous and inert particulate substrate.

Some authors (inter alia [3]) proposed simple procedures for the estimation of the biomass activity which is directly correlated with the number of the active cells present in a biomass. Among these procedures we can mention respirometric tests with oxygen utilization rate (OUR) measurements. Kappeler [4] and Gujer [5] proposed a respirometric method to estimate the heterotrophic active biomass, expressed in terms of COD. This value, inferred from OUR dynamics, is obtained by a test conducted during the exponential growth phase of bacteria fed with raw wastewater, and allows an indirect estimation of the active fraction in an activated sludge.

Hybridization techniques based on RNA staining with 16S- and 23S-rRNA-target fluorescent

*Corresponding author. Tel.: +39-0461-882-669; fax: +39-0461-882-672.
E-mail address: lisa@ing.unitn.it (P. Foladori).

0043-1354/02/$ - see front matter
PII: S0043-1354(01)00228-7

oligonucleotide probes are being increasingly utilized [6,7] and at present introduced to identify microorganisms in biofilm and in activated sludge plants. Even though these techniques allow the identification of specific bacteria species in the sludge, they do not permit to determine their activity.

The aim of this study is to investigate the possibility to directly estimate the dynamics of the microorganisms fraction of a sludge and the related kinetic parameters, as specific growth (μ) and decay (b) rate, utilizing flow cytometry (FCM) by means of direct single cell measurements. For this measure, it is essential to identify the individual cells within the activated sludge, to enumerate them, to estimate both viability and activity. Due to the diversity, heterogeneity and high number of cells that make up an activated sludge or a biofilm, these measurements require automation and speed. At present, the use of flow cytometry in conjunction with fluorescent viability dyes provides a promising method for the detection and rapid enumeration of viable bacteria in environmental samples [8–10] or sewage effluents [11]. An in-depth characterization of the so-called "viability concept" was proposed by Nebe-von-Caron et al. [12]. In a simple classification, the non-dead bacteria (dead bacteria are cells with permeabilized cytoplasmatic membrane, freely exposed to the environment and decomposed) are distinguished on the basis of their physiological status such as: (i) reproductive growth, (ii) metabolic activity, or (iii) membrane integrity [12]. The reproductive growth represents the most stringent proof for viability, because it requires both conditions (ii) and (iii). Cells with intact membrane are supposed to be capable of metabolic activity and show reproductive growth under appropriate conditions.

In this work, we utilized the terms "viable" bacteria and "active" bacteria to indicate cells showing intact membranes and metabolic activity, respectively. Fluorescent or fluorogenic dyes are frequently used as indicators for different cellular functions: (a) detection of membrane integrity using dye retention or dye exclusion [13]; (b) bacterial respiration using tetrazolium salts as INT or CTC [14]; (c) membrane potential using Oxonol or Rhodamine 123 [15], or (d) intracellular enzyme activity [8]. Among the (a) option, the membrane integrity analysis is based on the capacity of the cells to exclude fluorescent dyes, such as propidium iodide (PI) or ethidium bromide. The use of SYBR Green I coupled with PI has been recently proposed by Barbesti et al. [13]. SYBR Green I is a dye of a new generation of nucleic acid-specific stains such as TOTO, YOYO-1, PicoGreen or SYTO family, those have considerably improved the detection limit of FCM [16,17]. Fluorogenic esters can be successfully used to assess intracellular enzyme activity (option (b)). Examples include hydrolysis of fluorescein diacetate (FDA), carboxyfluorescein diacetate (CFDA) or 2',7'-bis-(2-carboxyethyl)-5-(and-6)-carboxyfluorescein acetoxy methyl ester (BCECF-AM). Fluorescent and/or fluorogenic dyes can be generally observed by a standard epifluorescence microscope (EFM), by means of a software for digital image analysis in the context of quantitative assessments. They have been recently implemented for flow cytometry analysis. EFM allows simple bacteria detection, but it is also laborious and time-consuming. A great advantage of FCM is the possibility to count an extremely great number of cells in a very short time—about 10^3 bacteria per second; therefore it is possible to obtain more precise values.

This work presents preliminary results obtained following the application of a rapid, automated method to assess the microbial viability and activity of sludge samples collected from wastewater treatment plants, coupling different techniques:

1. disaggregation of the biomass taken from the activated sludge oxidation tank;
2. identification of viable and dead cells by nucleic acids double staining with SYBR Green I + PI;
3. identification of enzymatic active and dead cells by double staining with FDA + PI and with BCECF-AM + PI;
4. FCM counting of the stained bacteria cells.

The results were applied to measure two important kinetic parameters. The calculated values were compared with those obtained from measurements of oxygen utilization rate (index of biomass activity) through conventional respirometric tests.

2. Materials and methods

2.1. Activated sludge samples

The viability measurements were carried out on biomass samples obtained from activated sludge MWWTPs in the province of Trento (Italy). Grab samples of mixed liquor have been collected from the oxidation tank and have immediately undergone a pre-treatment (within a few hours from collection). The pre-treatment aims to obtain a dispersion of the activated sludge flocs (usually their size reaches up to 200–300 μm) to separate the single bacteria.

2.2. Disaggregation of the sludge

Grab samples (150 mL) of activated sludge were preliminary homogenized with a mechanical blender (IKA-Turrax-T25; IKA Labortechnik, Staufen, Germany)—as reported by Wallner et al. [7]—for 3–5 min. A 5 mL volume of the homogenized samples was disaggregated with an IKA-Ultra-Turrax T8 set on 5 speed level for 10 min. Then the samples were diluted 1 : 16 (v/v) with 0.22-μm filtered phosphate-buffered saline solution (3 g K_2HPO_4, 1 g KH_2PO_4 and 8.5 g NaCl per liter; pH = 7.2). A further 10 min disaggregation step was carried out in order to reach the plateau of the bacteria counts. Fig. 1 describes the progressive recovery of viable and dead cells during pretreatment, determined through a double color DNA staining (SYBR Green I and PI, see further for the procedures) to investigate the effect of mechanical dispersion on the cell membrane integrity. The overall time of mechanical disaggregation should be of almost 15–20 min, in order to reach a plateau level of total cells (sum of both viable and dead). The count of viable and dead cells progressively increases from the beginning of the pre-treatment. If we consider the dead/viable cells ratio, a slight increase of this ratio appears starting after 15 min of mechanical disaggregation (see Fig. 1). This fact suggests a progressive damage and thus permeabilization of the viable cells. Therefore the pre-treatment was standardized to 4 min of homogenization, followed by 15 min of mechanical treatment in total. This period of mechanical action does not damage free cells as experimentally checked on bacterial strains. The application of sonication instead of mechanical dispersion did not yield satisfactory results. Protozoa in activated sludge samples are about $10^6\,mL^{-1}$ (three order of magnitude less than bacteria). Protozoa are destroyed during pre-treatment, as confirmed by numerous direct microscopic observations before and after treatment.

The pre-treated samples have to be diluted to give a final concentration of approximately 10^3 or 10^4 organisms per μL in order to be analyzed by FCM. In the case of activated sludge collected from Trento's MWWTPs the samples are diluted 150–300 times (v/v). A 10-μm pore-size filtration (Isopore membrane, Millipore) was finally done to avoid clogging of the cytometer nozzle. The aggregates eventually remaining on the membrane filter (residual small flocs, some dense cluster or filamentous cells) were stained with fluorescent dye (SYBR Green I and PI) and quantified at epifluorescence microscope by digital image acquisition system (CCD camera, Photometrics) and then compared with the flocs area in the same sludge sample before the pre-treatment. The residual flocs resulted lower than 3% (evaluated as area) of the initial stained biomass analyzed by image analysis. This fraction was lost and was thus not included in the cytometric quantification. This result testifies a good recovery of the whole procedure.

2.3. Fluorescent probes and staining conditions

The fluorochromes Propidium Iodide (PI) and SYBR Green I (Molecular Probes Inc., Eugene, Oregon) are widely reported as high-affinity nucleic acid dyes capable to give bright staining of the cells [13,18]. A 1 : 100 dilution of the SYBR Green I (referred to hereafter as SYBR I) commercial stock solution was made in dimethyl sulfoxide. The PI commercial stock solution has a concentration of $1\,mg\,mL^{-1}$. Staining with SYBR I and PI was made adding 10 μL of both fluorochromes for each mL of sample containing about 10^6–10^7 cells. The staining kinetics were determined by flow cytometry on samples of *E. coli* and samples of activated sludge. The double staining equilibrium was reached within 15 min of incubation at room temperature in the dark.

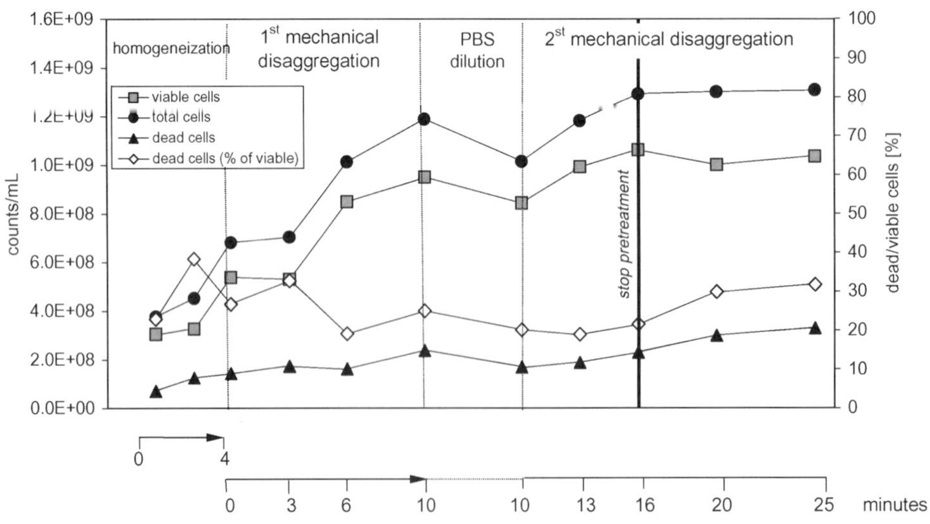

Fig. 1. Effect of the pre-treatment steps on the total, viable and dead bacteria recovery. The viable and dead bacteria determinations are based on membrane integrity.

Fluorogenic substrates as FDA (Sigma) and BCECF-AM (Molecular Probes, Eugene, Oregon) are used to evaluate enzymatic activity. 1 mM FDA and BCECF-AM solution were made with dimethylsulfoxide. 10 μL of a 0.01 mg mL^{-1} BCECF-AM solution and 10 μL of a 0.1 mM FDA solution should be used per mL of sample to obtain adequate signals. Samples were incubated for 30 min at 37°C in the dark, not washed and processed in few minutes.

2.4. Flow cytometer

A flow cytometer determines simultaneously two scattering signals (FALS, *Forward Angle Light Scatter* and LALS, *Low Angle Light Scatter*) and two fluorescence signals at specific wavelengths for each bacterial cell. In this study, samples were analyzed with a Bryte-HS flow cytometer (Bio-Rad, Hercules, CA), which is a purpose-built instrument for possible application in microbiology having a high light scattering sensitivity [19]. Bryte-HS is equipped with a short-arc xenon lamp (Hamamatsu, Japan). Samples were illuminated with an excitation of 470–490 nm and fluorescence emission was detected at 515–565 nm for FDA, BCECF and SYBR I ($\lambda_{ex} = 488$ nm, $\lambda_{em} = 525$ nm) and at 590–720 nm for PI ($\lambda_{ex} = 530$ nm, $\lambda_{em} = 620$ nm). Samples were analyzed for 3–4 min at a flow rate of 1.5 μL min^{-1} for cells enumeration. Data acquisition is triggered by green fluorescence to reduce interference by non-fluorescent particles (debris). Photomultiplier gains were set in the logarithmic mode (arbitrary scale expressed in channels, in the range 0–255) and data were recorded as listmode files using the WinBryte software (Bio-Rad). Graphs were drawn with the WinMDI 2.8 freeware (http://facs.scripps.edu/software.html). Combining the fluorescence and the scattering distribution measured by FCM single-cell analysis, it is possible to individuate functional subgroups of the bacteria population and to represent the results in 1-D histograms or a 2-D cytograms.

Sample observation was carried out utilizing a Nikon Labophot microscope at ×400 or ×1000 fitted for epifluorescence with a high-pressure mercury bulb and filter set type UV-2A EX 330–380 and G2A EX510-560.

2.5. OUR measurement apparatus

The OUR tests were carried out to monitor the respirometric activity of the biomass. A 2-liters controlled temperature and aerated reactor was utilized. Mixing is guaranteed by a magnetic stirrer at a revolution speed that avoids mixed liquor re-oxygenation. Dissolved oxygen was monitored by an oxymeter (OXI 340, WTW GmbH, Weilheim, Germany) connected to a data acquisition system. Fig. 2 shows the arrangement of the instrumentation used. The VSS and TSS were measured to conventionally characterize activated sludge biomass according to Standard Methods [20].

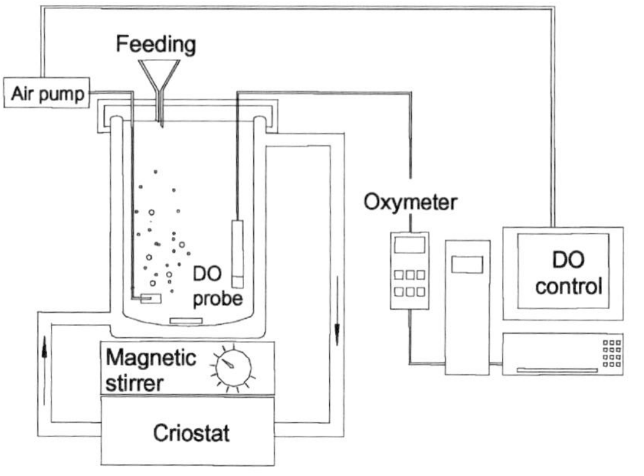

Fig. 2. Scheme of the instrumentation utilized for the respirometric runs.

3. Results and discussion

3.1. Cytometric analysis

The staining combinations described in the previous paragraphs have been applied to activated sludge samples. In Sections 3.1.1–3.1.4, the results obtained by FCM analysis with the different dyes are described.

3.1.1. Membrane integrity as indicator of viability

The determination of viable and dead bacterial cells in activated sludge samples was done on the basis of membrane integrity coupling SYBR I and PI. The SYBR I is a molecule capable of staining all cells, whether living or dead. No specific information is available on its structure and chemical properties and in particular on its binding mode to DNA. The PI is capable of staining only dead or damaged cells (because of its polarity, PI penetrates only inactive cells membranes). In dead cells the simultaneous presence of SYBR I and PI activates the so-called "energy transfer" phenomenon, i.e., the fluorescent emission spectrum of SYBR I is absorbed by PI and no longer visible. Thus, it is possible to distinguish viable cells (green fluorescence) from dead ones (red fluorescence). In damaged cells (neither dead nor viable) the energy transfer is not complete because of the lower PI concentration; damaged cells emit both green and red fluorescences. The flow cytometer produces a red versus green fluorescence cytogram in which single bacteria are

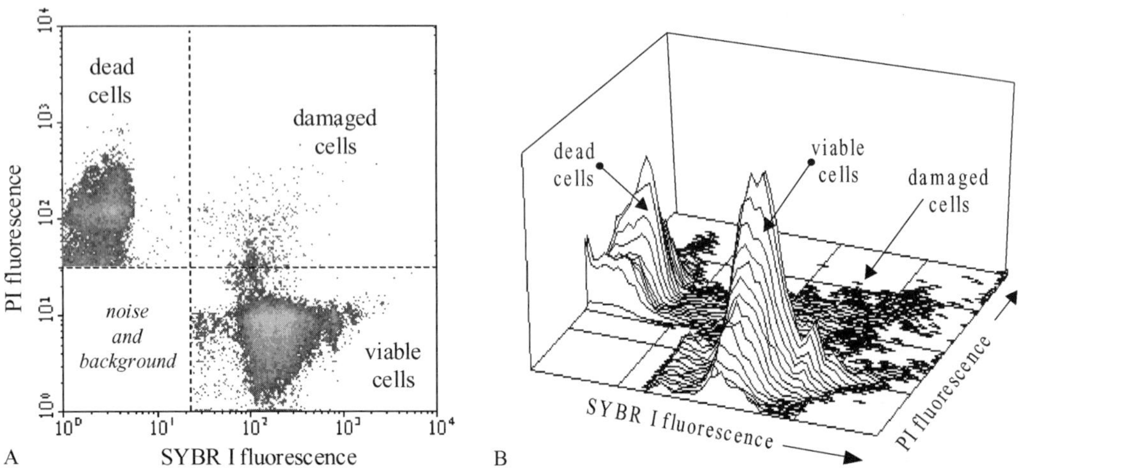

Fig. 3. Red versus green fluorescence cytogram for activated sludge cells stained with SYBR I + PI. The diagrams refer to a total count of about 20,000 cells: (A) 2-D distribution (cell density in each fluorescence channel proportional to dot plot density); (B) 3-D distribution.

distributed accordingly (see Fig. 3 (A)). In the cytogram it is possible to distinguish four areas: dead bacteria, viable bacteria, damaged bacteria and noise (that includes the particles with low fluorescence in both wavelengths and the instrument background). With a 3-D representation (see Fig. 3 (B)) it is possible to visualize simultaneously the two fluorescence signals on the horizontal plane, while on the vertical axis the number of events for a couple of coordinates on the horizontal plane are shown. Fluorescence obtained by SYBR + PI is stable and therefore allows contemporaneous staining of few samples.

3.1.2. Enzymatic activity assessment

Enzymatic activity was determined by double staining with fluorogenic BCECF-AM and PI and double staining with fluorogenic FDA and PI [10]. When taken up by active cells, the non-fluorescent FDA and BCECF-AM are hydrolyzed by intracellular non-specific esterases to produce fluorescein and BCECF, respectively, that are fluorescent and retained by cells having intact membrane. The fluorescence produced by BCECF-AM and FDA staining tends to substantially decrease in few minutes after the incubation.

3.1.3. Comparison among different staining procedures

Fig. 4 shows the red and green fluorescence distribution and scattering signals (in logarithmic scale) of cells obtained from activated sludge samples stained with the three combinations of dyes tested. Forward light scattering is correlated to cell size. Preliminary observations show that the typical size of the single cells are in the range 0.2–2 μm. A good separation of the viable/active and the dead populations of cells was observed in all the cases. Dead cells are approximately 22% (on average) of the total cells counted. Best distinction signal to noise on the green fluorescence was obtained with SYBR I and BCECF-AM (see Fig. 4). The fluorescent noise while staining with FDA interferes with the stained bacteria counts, probably because of the presence of unspecific staining of cellular debris or particulate matter. This interference, producing a partial overlapping of the signal and noise fluorescence distributions, can result in somewhat biased estimations of bacterial counts. In the analyzed samples this interference causes errors in the estimation of the active cells counts with FDA staining. This dye was not chosen in the further application to full-scale plants. The numbers of viable bacteria (determined by SYBR I + PI) present in the treatment plant were always higher than those enzymatically active (counted with BCECF + PI). The ratio between active and viable bacteria was about 0.8, while the ratio between active and total bacteria was about 62%. Yamaguchi and Nasu [10] found similar values (50–90%) in river water bacteria population.

3.1.4. Accuracy and repeatability of the FCM count

Performances of the whole method (pre-treatment, staining and FCM analysis) were tested using SYBR I and PI dyes. Accuracy was tested comparing five cytometric viable counts in the usual working range of 5000–10,000 cells per μL against standard cell counts with epifluorescence microscope and Bürker chamber. The accuracy was 94% on average, ranging from 86% to 110%.

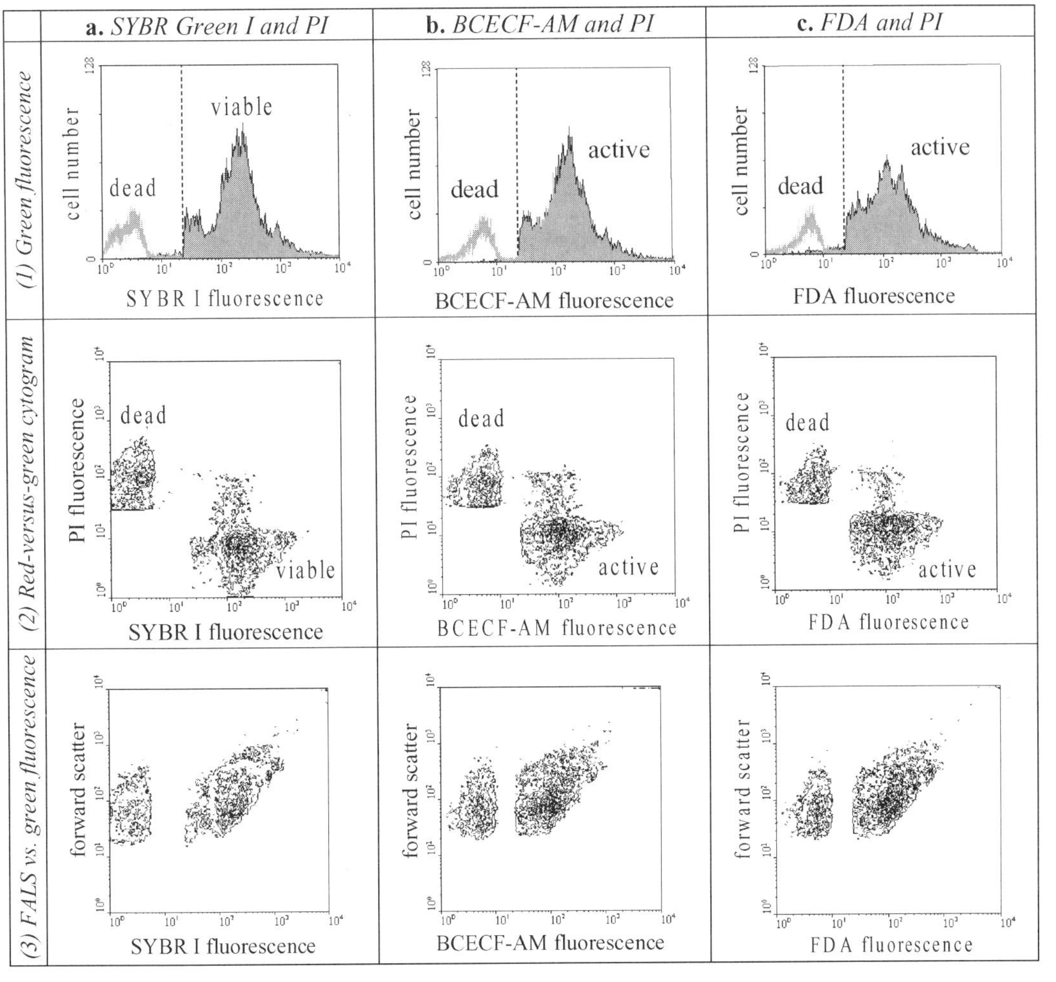

Fig. 4. Red and green fluorescence and forward light scattering distributions of bacteria from activated sludge samples. The columns refer to the different staining: (a) SYBR Green I + PI; (b) BCECF-AM + PI and (c) FDA + PI. The line refers to: (1) green fluorescence histograms of viable and dead cells; (2) dot plot of red versus green fluorescence and (3) dot plots of forward angle light scattering (FALS) versus green fluorescence.

Repeatability of the FCM count was assessed calculating CV (coefficient of variation) on five samples of the same mixed liquor. A 2 L mixed liquor sample was split into five 100 mL independent subsamples; each subsample was pretreated and analyzed by cytometry as three replicates. The intra-sample CV for total and viable bacteria (three replicate cytometric analysis on each pre-treated subsample) did not outreach 10% (range 2.4–9.7%). Inter-subsamples CV did not outreach 12% of the total bacteria (range 8.4–11.7%) and 16% of the viable bacteria (range 14.5–16.0%). The inter-samples variability is higher than the intra-sample one, suggesting that the pre-treatment procedure used to obtain a single-cell suspension plays a residual role. We repeated this procedure three times, always obtaining the same figures.

4. Applications to an activated sludge WWMTP

Cytometric analysis performed on activated sludge samples collected at the Trento Nord WWTP (100.000 PE, sludge age about 17 days and F/M ratio of 0.13 $kgBOD_5\,kgSST^{-1}\,d^{-1}$) over several months period yielded the following average results: 1.24E+09 viable bacteria counts mL^{-1} (standard deviation 0.24E+09), 4.17E+08 dead bacteria counts mL^{-1} (standard deviation 1.57E+08), 2.38E+08 damaged bacteria counts mL^{-1} (standard deviation 0.64E+08). The average concentrations of TSS in the analyzed samples was equal to 3.5 $kgSST\,m^{-3}$ (standard deviation. 0.6) with a 0.7 VSS/TSS ratio. Utilizing plate count agar (incubation at 22°C for 7 days), it was only possible to count 1.5% of the bacteria measured with FCM. This does agree with other authors' findings: they report values up to 5–15% of the total bacteria in spite of the optimization of the cultural medium [6].

Viability and activity measurements were applied in this study to measure the two main kinetic parameters in the biological activated sludge processes:

(1) The specific endogenous decay rate, determined in a no feed batch test.
(2) The specific growth rate, determined through an exponential growth curve in the presence of a high *Food/Microorganism* ratio (S/X ratio).

4.1. Endogenous decay rate

A 7-days batch test was performed to estimate endogenous decay rate, using a 30-L mixed liquor continuously aerated, collected from the oxidation tank of an activated sludge plant and kept under endogenous conditions. OUR tests, viable and active cells counts (with SYBR I+PI, BCECF-AM+PI and FDA+PI), TSS and VSS analysis were daily performed. Respirometric tests have been carried out adding to a 1.5 L of mixed liquor 100–150 mg of sodium acetate as RBCOD and nitrification inhibitor (ATU), according to the method proposed by Spanjers and Vanrolleghem [21].

Test results are shown graphically in Fig. 5. The decay coefficient may be calculated from the slope of the plot of the natural logarithm of each measured parameter versus time. Therefore we can estimated four decay rates, as follows: 0.116 d^{-1} (from OUR), 0.105 d^{-1} (from viable cells), 0.117 d^{-1} (from active cells) and 0.015 d^{-1} (from VSS). These preliminary observations are close to data referred for activated sludge, based on respirometric measurements: $0.092 \pm 0.051\ d^{-1}$ [22], 0.04 d^{-1} [21] and 0.09 d^{-1} [23]. OUR activity decay in endogenous phase was proportionally related with the decay of both viable and active bacteria. We could conclude about the good accuracy of the simple OUR in estimating this constant. The ratio between enzymatically active and viable bacteria was 0.85 on average (range 0.7–0.95). VSS decrease was very low and did not represent the real biomass evolution during the test, being biased by the fate of the biodegradable particulate and the presence of inert.

4.2. Specific growth rate

Kappeler and Gujer [4] proposed a batch test based on OUR measurement for assessing the specific growth rate. The test needs to be carried out with a high S/X ratio, allowing an unlimited heterotrophic bacterial growth. As soon as the readily biodegradable COD is exhausted, OUR immediately decreases. About 20 mL of mixed liquor having a concentration of 3.5 gTSS L^{-1}

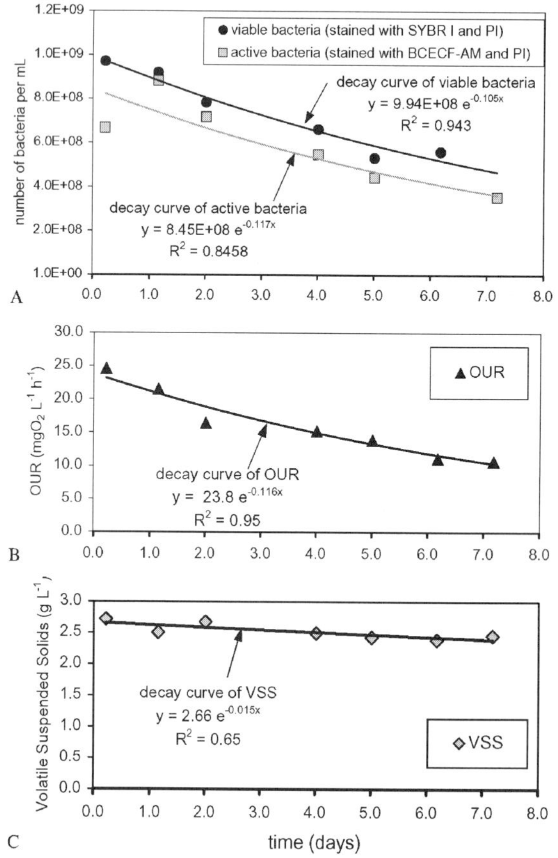

Fig. 5. Experimental data and exponential fitting curves relative to the endogenous decay test: (A) viable and active bacteria dynamics; (B) OUR dynamics and (C) VSS dynamics.

were added to a centrifuged raw wastewater sample (1.5 L). Additional Sodium Acetate was added to obtain an above $4S/X$ ratio. The respirometric test lasted about 11 h. Samples for cytometric analysis were collected at the beginning, during the growth and at the end of the exponential growth phase. Viable counts were measured to estimate the net growth rate. The initial number of viable bacteria (7.3E+05 cells per mL) increased exponentially to the final concentration of 2.3E+06 cells per mL. Results are shown in Fig. 6. The kinetic constants μ_{max} can be calculated from the OUR, according to Kappeler and Gujer, and can be compared with the direct measure of the viable cells increase. The exponential fitting curve plotted in Fig. 6 allows to calculate ($\mu_{max} - b_H$) for each set of data. Knowing b_H (see previous paragraph) μ_{max} can be directly calculated as 6.24 d^{-1} (OUR data) and 3.4 d^{-1} (viable cells). Ekama et al. [24] and Kappeler and Gujer [4] calculated values of 6 and 8 d^{-1}, respectively, by respirometric techniques. In this paper, only the number of viable (and active) cells

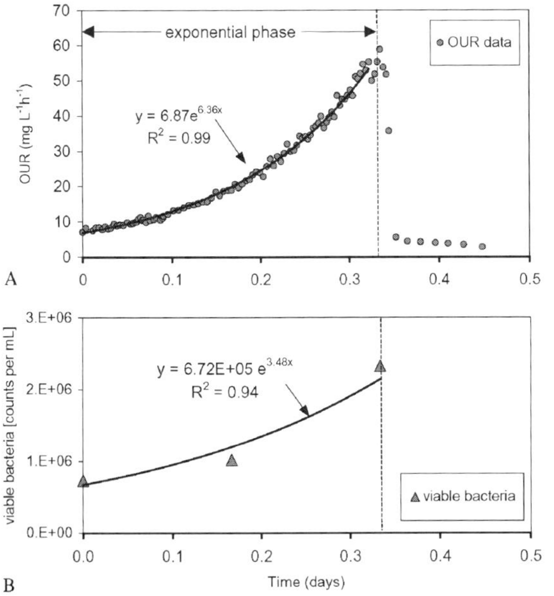

Fig. 6. (A) OUR dynamics according to Kappeler and Gujer [9]; (B) cytometric analysis results.

were measured and not the viable mass on the basis of biovolume. Therefore the estimated value needs to be further confirmed.

5. Conclusions

The aim of this study was to demonstrate the potential of fluorescent dyes combined with FCM as tools for the viability and activity assessment of an activated sludge bacterial mixed population. A procedure has been developed to disaggregate sludge flocs before dyes staining and cytometric analysis. The developed procedure allows a high recovery of bacteria with good accuracy and repeatability, and minimize the damage of the cells suspension obtained from the disaggregation of the flocs. Double staining with SYBR Green I and propidium iodide was chosen to detect the cellular membrane integrity and was used to count the viable dispersed bacteria, while double staining with the fluorogenic dye BCECF–AM and propidium iodide was chosen to count the enzymatically active bacteria. The ratio active/viable bacteria resulted similar for activated sludge samples in steady-state conditions and in endogenous phase (prolonged to a week, period similar to the sludge retention time in a waste water treatment plant); the average value was quite constant (0.8). Flow cytometry, widely used in clinical and recently in the environmental field, also appear to be a very promising tool to investigate bacteria dynamics on wastewater treatment plants. Preliminary results on growth and decay rates, are encouraging and support this possibility.

Acknowledgements

This project is financially supported by the Province of Trento. The authors owe a special thanks to Mrs. A. Gelmini for her collaboration on performing respirometric tests. The authors appreciated the comments of the anonymous reviewers.

References

[1] Gujer W, Henze M, Mino T, Matsuo T, Wentzel MC, Marais GVR. Activated sludge model No. 2, biological phosphorus removal. Water Sci Technol 1995;31(2):1–11.

[2] Henze M, Grady Jr CPL, Gujer W, Marais GVR, Matsuo T. IAWPRC Task Group on mathematical modelling for design and operation of biological wastewater treatment. Activated Sludge Model No. 1, IAWQ Scientific and Technical Reports No. 1. IAWPRC, London, UK, 1987.

[3] Jørgensen PE, Eriksen T, Jensen BK. Estimation of viable biomass in wastewater and activated sludge by determination of ATP, oxygen utilization rate and FDA hydrolysis. Water Res 1992;26(11):1495–501.

[4] Kappeler J, Gujer W. Estimation of kinetic parameters of heterotrophic biomass under aerobic conditions and characterization of wastewater for activated sludge modelling. Water Sci Technol 1992;25(6):125–39.

[5] Wentzel MC, Ubisi MF, Ekama GA. Heterotrophic active biomass component of activated sludge mixed liquor. Water Sci Technol 1998;37(4/5):79–87.

[6] Wagner M, Amann R, Lemmer H, Schleifer KH. Probing activated sludge with oligonucleotides specific for proteobacteria: inadequacy of culture-dependent methods for describing microbial community structure. Appl Environ Microbiol 1993;59:1520–5.

[7] Wallner G, Erhart R, Amann R. Flow cytometric analysis of activated sludge with rRNA-targeted probes. Appl Environ Microbiol 1995;61(5):1859–66.

[8] Diaper JP, Edwards C. The use of fluorogenic esters to detect viable bacteria by flow cytometry. J Appl Bacteriol 1994;77:221–8.

[9] Porter J, Deere D, Hardman M, Edwards C, Pickup R. Go with the flow—use of flow cytometry in environmental microbiology. FEMS Microbiol Ecol 1997;24:93–101.

[10] Yamaguchi N, Nasu M. Flow cytometric analysis of bacterial respiratory and enzymatic activity in the natural aquatic environment. J Appl Bacteriol 1997;83:43–52.

[11] Porter J, Edwards C, Morgan JAW, Pickup RW. Rapid, automated separation of specific bacteria from lake water and sewage by flow flow cytometry and cell sorting. Appl Environ Microbiol 1993;59(10):3327–33.

[12] Nebe-von-Caron G, Stephens PJ, Hewitt CJ, Powell JR, Badley RA. Analysis of bacterial function by multi-colour fluorescence flow cytometry and single cell sorting. J Microbiol Meth 2000;42:97–114.

[13] Barbesti S, Citterio S, Labra M, Baroni MD, Neri MG, Sgorbati S. Two and three-color fluorescence flow cytometric analysis of immunoidentified viable bacteria. Cytometry 2000;10:214–8.

[14] López-Amorós R, Comas J, García MT, Vives-Rego J. Use of the 5-cyano-2,3-ditolyl tetrazolium chloride reduction test to assess respiring marine bacteria and grazing effects by flow cytometry during linear alkylbenzene sulfonate degradation. FEMS Microbiol Ecol 1998;27:33–42.

[15] Kaprelyants AS, Kell DB. Dormancy in stationary-phase cultures of Micrococcus luteus: flow cytometric analysis of starvation and resuscitation. Appl Environ Microbiol 1993;59(10):3187–96.

[16] Lebaron P, Parthuisot N, Catala P. Comparison of blue nucleic acid dyes for flow cytometric enumeration of bacteria in aquatic systems. Appl Environ Microbiol 1998;64(5):1725–30.

[17] Marie D, Partensky F, Jacquet S, Vaulot D. Enumeration and cell cycle analysis of natural populations of marine picoplankton by flow cytometry using the nucleic acid stain SYBR Green I. Appl Environ Microbiol 1997;63(1):186–93.

[18] Weinbauer MG, Beckmann C, Höfle MG. Utility of green fluorescent nucleic acid dyes and aluminum oxide membrane filters for rapid epifluorescence enumeration of soil and sediment bacteria. Appl Environ Microbiol 1998;64(12):5000–3.

[19] Steen HB, Lindmo T. Flow cytometry: a high resolution instrument for everyone. Science 1979;204:403–4.

[20] APHA. Standard methods for the examination of water and wastewater. APHA, AWWA and WPCF, Washington, DC, USA, 1995.

[21] Spanjers H, Vanrolleghem P. Respirometry as a tool for rapid characterization of wastewater and activated sludge. Water Sci Technol 1995;31(2):105–14.

[22] Keesman KJ, Spanjers H, van Straten G. Analysis of endogenous process behaviour. Water Sci Technol 1998;37(12):227–35.

[23] Avcioglu E, Orhon D, Sözen S. A new method for the assessment of heterotrophic endogenous respiration rate under aerobic and anoxic conditions. Water Sci Technol 1998;38(8/9):95–103.

[24] Ekama GA, Siebritz IP, Marais GvR. Considerations in the process design of nutrient removal activated sludge processes. Water Sci Technol 1983;15(3/4):283–318.

PERGAMON

Water Research 36 (2002) 469–481

WATER RESEARCH

www.elsevier.com/locate/watres

Nitrifying and heterotrophic population dynamics in biofilm reactors: effects of hydraulic retention time and the presence of organic carbon

Regina Nogueira[a,b], Luís F. Melo[a], Ulrike Purkhold[b], Stefan Wuertz[c], Michael Wagner[b,*]

[a] Department of Chemical Engineering, Faculty of Engineering of Porto, LEPAE, Porto, Portugal
[b] Lehrstuhl für Mikrobiologie, Technische Universität München, Am Hochanger 4, D-85350 Freising, Germany
[c] Technische Universität München, Am Coulombwall, D-85748 Garching, Germany

Received 29 September 2000; received in revised form 09 January 2001; accepted 1 February 2001

Abstract

Two biofilm reactors operated with hydraulic retention times of 0.8 and 5.0 h were used to study the links between population dynamics and reactor operation performance during a shift in process operation from pure nitrification to combined nitrification and organic carbon removal. The ammonium and the organic carbon loads were identical for both reactors. The composition and dynamics of the microbial consortia were quantified by fluorescence in situ hybridization (FISH) with rRNA-targeted oligonucleotide probes combined with confocal laser scanning microscopy, and digital image analysis. In contrast to past research, after addition of acetate as organic carbon nitrification performance decreased more drastically in the reactor with longer hydraulic retention time. FISH analysis showed that this effect was caused by the unexpected formation of a heterotrophic microorganism layer on top of the nitrifying biofilm that limited nitrifiers oxygen supply. Our results demonstrate that extension of the hydraulic retention time might be insufficient to improve combined nitrification and organic carbon removal in biofilm reactors.

Keywords: Nitrification; Biofilm; Organic carbon; Hydraulic retention time; Fluorescence in situ hybridization

1. Introduction

The competition between heterotrophic and nitrifying bacteria for substrates (oxygen and ammonia) and space in biofilms is of major practical importance and thus has been the subject of several previous studies (for example [1–3]). According to these investigations, competition in biofilms results in a stratified biofilm structure, the fast growing heterotrophic bacteria being placed in the outer layers, where both substrate concentration and detachment rate are high, while the slow growing nitrifying bacteria stay deeper inside the biofilm. Thus a heterotrophic layer can form above the nitrifiers in the biofilm, which constitutes a disadvantage to them when the bulk liquid oxygen concentration is low. In this case oxygen limitation resulting from consumption and resistance to mass transfer within the heterotrophic layer affects the nitrification performance negatively. As long as the bulk oxygen concentration is high enough to preclude its depletion in the biofilm, however, the heterotrophic layer can also have a positive effect on the nitrifiers by protecting them from detachment [4].

One possible approach to minimize the competition of heterotrophs and nitrifiers for oxygen is their spatial separation into a nitrifying biofilm population and a

*Corresponding author. Tel.: +49-8161-71-5444; fax: +49-8161-71-5475.

E-mail address: wagner@mikro.biologie.tu-muenchen.de (M. Wagner).

0043-1354/02/$ - see front matter
PII: S0043-1354(01)00229-9

heterotrophic population in suspension. This separation can be achieved by extending the reactor's hydraulic retention time. At comparatively long hydraulic retention times, the fast growing heterotrophic microorganisms (with reciprocal maximum specific growth rates smaller than the selected hydraulic retention time) grow mainly in suspension while the slow growing nitrifiers form biofilms [2]. However, in this study the reactor was operated at a high oxygen concentration of $6\,mg\,l^{-1}$, a condition which did also allow full nitrification in a conventional biofilm reactor (nitrifiers and heterotrophs in the biofilm) [2].

Reactors with a spatial separation of heterotrophic and nitrifying activity for the treatment of effluents with a high chemical oxygen demand (COD)/NH_4^+-N ratio can be very attractive for future practical applications if (i) it can be demonstrated that these reactors allow complete nitrification at a relatively low dissolved oxygen concentration and if (ii) the energy savings in aeration relatively to the traditional process compensate the investment costs in the construction of a bigger reactor required due to its longer retention time. So far the effects of influent composition and operational conditions were studied with respect either to total reactor performance (macroscale studies) or to biofilm composition and structure (microscale studies), but hardly to both simultaneously (for example [3]).

The research subjects of this study were the effects of different hydraulic retention times and changes in the organic carbon dosing on the population dynamics of nitrifying biofilm reactors operated at around $2\,mg\,l^{-1}$ dissolved oxygen. The specific objectives were: (i) to identify and quantify nitrifying and heterotrophic bacteria in the biofilm and in suspension using a set of rRNA-targeted oligonucleotide probes for fluorescence in situ hybridization (FISH), (ii) to correlate changes in microbial community composition in the biofilm and suspension with reactor performance during a shift from pure nitrification to combined nitrification and organic carbon removal, and (iii) to asses the recovery of the nitrification process after a shift back to pure nitrification.

2. Materials and methods

2.1. Biofilm reactors

Two laboratory-scale circulating bed reactors (CBR) of 1.2 l each [5] were employed for this study. This type of airlift reactor has a rectangular geometry of 310 mm height and 60 mm × 60 mm cross-section and is separated in an up-flow aerated section and a down-flow non-aerated one by a vertical wall (Fig. 1). High density polyethylene granulate with a particle size of 1 mm and a density of $731\,kg\,m^{-3}$ was used as support material for biofilm growth. The superficial air velocity in both

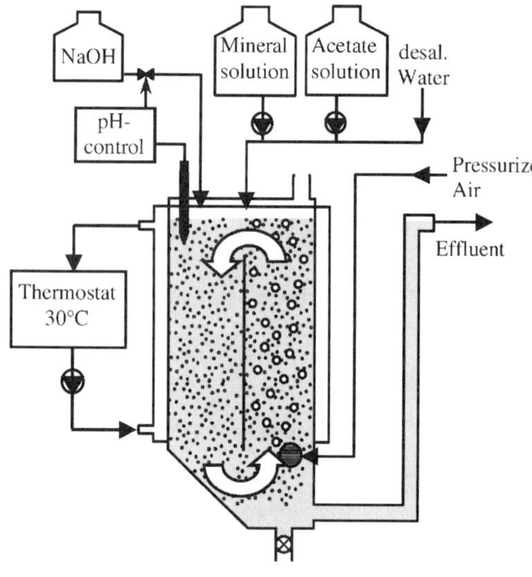

Fig. 1. Schematic diagram of continuous circulating bed reactor set-up.

reactors (defined as the air flow rate divided by the aerated reactor cross-section) was set at $0.003\,m\,s^{-1}$. The temperature was maintained at 30°C, and the pH was kept at 7.5 by adding sodium hydroxide (1 M). The experimental protocol included 4 phases (Fig. 2). The corresponding experimental conditions are summarized in Table 1.

Phase I. A CBR (hereafter reactor R_0) was filled with nitrifying biofilm particles (23 volume percent) obtained from a nitrifying circulating bed reactor, which was maintained at identical operating conditions. The biofilm particles had been stored at 4°C for 90 days prior to inoculation of reactor R_0. An ammonium solution was supplied continuously (N operation mode), with a retention time of 0.70 h. During phase I a stable nitrifying biofilm was established.

Phase II. Half of the biofilm particles from reactor R_0 were transferred to a second identical reactor. Both CBRs (hereafter reactors R_1 and R_2) were operated simultaneously with retention times of 0.8 h for reactor R_1 and 5.0 h for reactor R_2. The ammonium load supplied to each reactor was approximately half of the value used in phase I in order to maintain a constant NH_4^+-N load to biomass ratio ($0.46\,kg\,kg^{-1}\,d^{-1}$), and still no organic carbon was added (N operation mode).

Phase III. Acetate as an organic carbon source was supplied to reactors R_1 and R_2 at the same COD/NH_4^+-N mass ratio (C+N operation mode) in order to investigate the effect of organic carbon on the nitrification performance of reactors operating at different retention times.

Phase IV. Discontinuation of the supply of acetate to reactors R_1 and R_2 in order to study the recovery of the nitrification process (N operation mode).

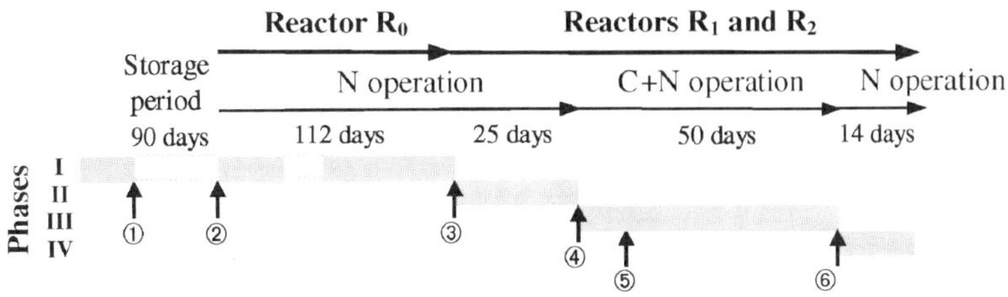

Fig. 2. Schematic diagram describing the experimental phases. The arrows indicate biofilm and suspended biomass sampling times.

Table 1
Operating conditions and performance of reactors R_0, R_1, and R_2. For all parameters average values are given for the different phases of operation

Phase/ mode of operation		Influent		NH_4^+-N load		NH_4^+-N removal (%)	COD load		COD removal (%)	COD/N applied ($g\,g^{-1}$)
		NH_4^+-N	COD ($kg\,m^{-3}$)	Applied	Removed ($kg\,m^{-3}\,d^{-1}$)		Applied	Removed ($kg\,m^{-3}\,d^{-1}$)		
Reactor R_0 hydraulic retention time of 0.7 h										
I	N	0.037	—[a]	1.26	1.20	95	—	—	—	—
Reactor R_1 hydraulic retention time of 0.8 h										
II	N	0.020	0.01	0.61	0.60	98	0.33	0.04	12	0.53
III	C+N	0.022	0.03	0.65	0.45	69	0.85	0.61	72	1.30
IV	N	0.019	0.01	0.57	0.51	90	0.17	0.05	29	0.30
Reactor R_2 hydraulic retention time of 5.0 h										
II	N	0.100	0.02	0.48	0.48	100	0.10	0.04	40	0.21
III	C+N	0.109	0.14	0.53	0.53→0	100→0	0.68	0.61	90	1.28
IV	N	0.117	0.02	0.56	0.53	94	0.08	0.05	63	0.14

[a] — = not determined.

Phases I, II and IV are characteristic of a pure nitrification process, while phase III corresponds to a combined nitrification and organic carbon removal process.

2.2. Media

During the experimental phases I, II and IV (N operation mode) 3.4 $ml\,min^{-1}$ of an ammonium medium was delivered to each reactor. The medium contained $(NH_4)_2SO_4$ (0.59 $g\,l^{-1}$), $NaHCO_3$ (0.19 $g\,l^{-1}$), KH_2PO_4 (0.06 $g\,l^{-1}$), $CaCl_2\,2H_2O$ (0.014 $g\,l^{-1}$) and trace elements. In reactors R_0 and R_1, also 15.9 $ml\,min^{-1}$ of deionized water was added to the reactors to obtain the desired hydraulic retention time. During phase III (C+N operation mode) 1.7 $ml\,min^{-1}$ of a sterilized acetate solution ($CH_3COONa\,3H_2O$, 0.71 $g\,l^{-1}$) was added separately to reactors R_1 and R_2. In order to maintain the retention time in both reactors constant, the ammonium medium's volumetric flow rate was reduced to half, and its concentration was doubled.

2.3. Overall kinetics

The macroscale reactor performance was evaluated from effluent samples filtered with 0.22 μm membrane filters. COD, ammonia plus ammonium, nitrite and nitrate were determined photometrically (LCK, Dr. Lange). Biofilm and suspended biomass total solids were measured according to APHA [6] using 0.22 μm membrane filters. Prior to this analysis, the biofilm was detached from the support material by an ultrasonic homogenizer (Bandelin electronics D-1000, Berlin) treatment (120 s at 50 W). The dissolved oxygen concentration was measured with an oxygen electrode (WTW, model Oxi 340-A).

2.4. Fluorescence in situ hybridization, microscopy and quantification of probe-targeted bacteria

Microbial population dynamics in biofilm particles and suspended biomass was evaluated using FISH with rRNA-targeted oligonucleotide probes. Samples 1–6

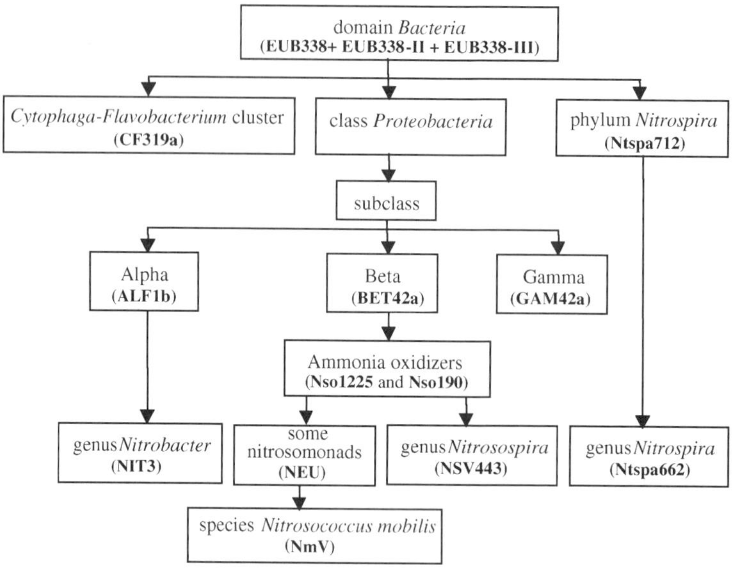

Fig. 3. Specificity of the rRNA-targeted oligonucleotide probes used for in situ identification of nitrifiers and heterotrophic bacteria.

(Fig. 2) were taken from the reactors and immediately fixed with paraformaldehyde. During phases I and II (N operation) only biofilm was sampled since the suspended biomass concentration was very low, while during phase III (acetate addition) both biofilm and suspended biomass samples were taken.

In situ characterization of microbial populations followed a top to bottom approach (Fig. 3). First the samples were hybridized with a probe set (EUB338, EUB338-II, EUB338-III) designed to target almost all bacteria [7], then with previously published group specific probes (Fig. 3; [8,9]). The ammonia-oxidizing and nitrite-oxidizing bacteria were identified using previously published probes (Fig. 3; [10–14]).

Oligonucleotide probes were purchased as derivatives labeled with the fluorescent dyes Cy3, Cy5, and 5(6)-carboxyfluorescein-N-hydroxysuccinimide-ester (FLUOS), respectively (Interactiva, Ulm, Germany). FISH was performed using the hybridization and washing buffers as described by Manz et al. [8]. A Zeiss LSM 510 laser scanning confocal microscope (Zeiss, Jena, Germany) was used for image acquisition. For quantification of probe-targeted bacteria, simultaneous hybridizations were performed with Cy3 labeled specific probes and the FLUOS labeled bacterial probe set. The relative biovolume defined as the ratio between the area of probe-targeted bacteria to the area of all bacteria detectable by FISH was determined for each probe in 20 randomly recorded confocal images (thickness 1 μm) using the procedure described by Schmid et al. [15]. Biofilm thickness was determined for fresh, unfixed biofilm samples which were stained with a $0.25\,g\,l^{-1}$ fluorescein isothiocianate solution for 3 h at room temperature, using CLSM optical sectioning in the sagittal (*xz*) direction.

2.5. Comparative sequence analyses of the *amoA* gene

High resolution analyses of ammonia-oxidizer diversity in reactor R_0 was performed using the gene encoding the catalytic subunit of the ammonia-monooxygenase enzyme (*amoA*) as a marker. Amplification, cloning, sequencing and phylogenetic analyses of the biofilm-derived *amoA* fragments was performed as described by Purkhold et al. [16].

3. Results

3.1. Reactor performance

3.1.1. Phase I

One of the main drawbacks of biological nitrification processes is the requirement of a long start-up period. By using biofilm particles as inoculum, the start-up period of reactor R_0 in phase I could be reduced to 4 days after which the ammonium removal efficiency had reached 95%. Subsequently, R_0's biofilm particles were split between reactors R_1 and R_2 to ensure that both reactors had the same original microbial population composition.

3.1.2. Phases II to IV

Fig. 4 (A–D) depicts the performance of reactors R_1 and R_2 operated with retention times of 0.8 and 5.0 h, respectively, during pure nitrification (phases II and IV) and combined nitrification and organic carbon removal (phase III). Tables 1 and 2 summarize the experimental results obtained during reactors operation and the characterization of biofilm and suspended biomass. Though acetate as organic carbon source was only added during phase III, there was a certain background COD in the influent during phases I, II and IV (Table 1; Figs. 4B and D) deriving from oxidizable matter in the deionized water source. However, during these phases maximal 63% of the incoming COD was removed demonstrating that a significant fraction of these compounds were not degraded in the reactors.

During phase II, both reactors had a stable performance, no nitrite accumulation was observed and the NH_4^+-N effluent concentration was below $1.0\,mg\,l^{-1}$ corresponding to an ammonium removal efficiency higher than 95%. The biofilm characterization at the end of phase II showed similar biofilm thickness (41 and 42 μm), and biofilm mass concentration (2.48 and $2.43\,kg\,m^{-3}$) in reactors R_1 and R_2.

Shortly after the addition of acetate to reactor R_1 (start of phase III), the ammonium removal rate decreased from 0.65 to $0.45\,kg\,m^{-3}\,d^{-1}$ (69%) and hereafter, was constant (Fig. 4A). Due to a mechanical problem in the ammonium dosing pump, reactor R_1 received an ammonium overload 10 days after the start of phase III (gray area in Fig. 4A). The ammonium removal rate, however, remained constant. In reactor R_2

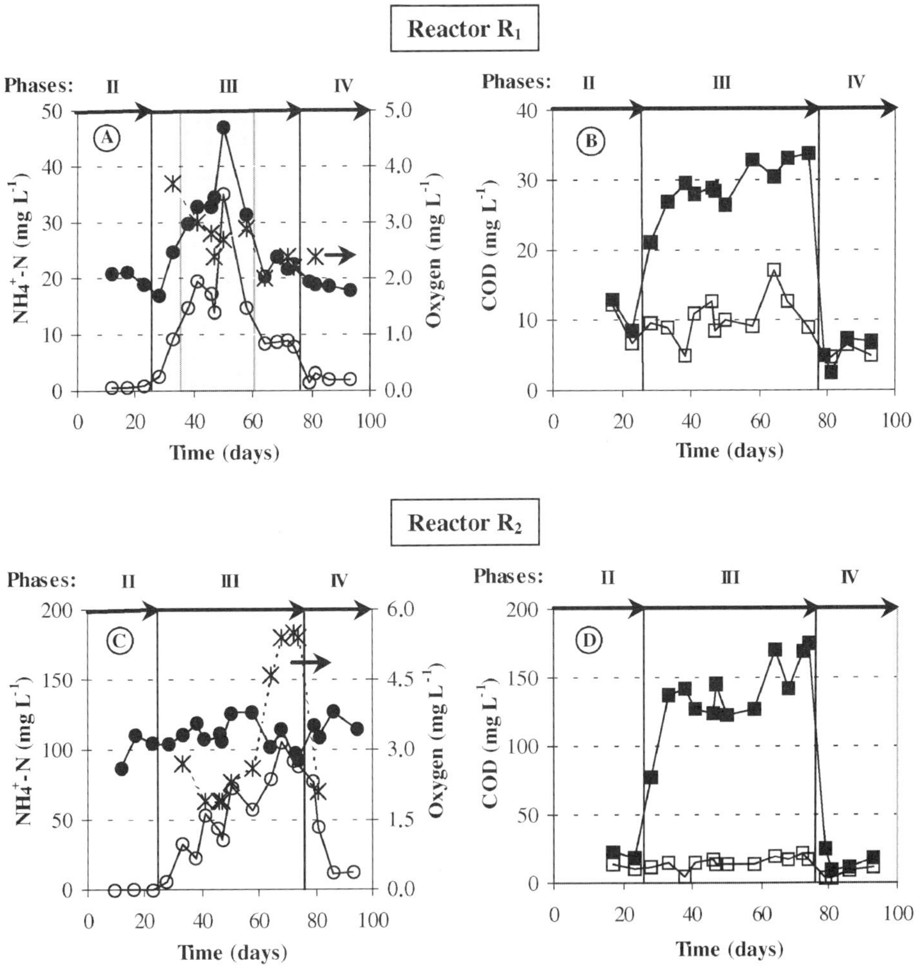

Fig. 4. Time changes of NH_4^+-N, COD and oxygen during operation of reactors R_1 and R_2 with pure nitrification (phases II and IV) and combined nitrification and organic carbon removal (phase III). Closed symbols correspond to influent concentrations and open symbols to effluent concentrations. *Corresponds to oxygen concentrations.

Table 2
Characterization of biofilm and suspended biomass in reactors R_0, R_1, and R_2 during the different phases of operation. Values listed in the table are the average ±95% confidence interval

Mode of operation/ sample		Biofilm mass per reactor volume ($kg\,m^{-3}$)	Suspended solids concentration ($kg\,m^{-3}$)	Biofilm thickness (μm)
Reactor R_0 hydraulic retention time of 0.7 h				
N	3	2.70±0.80	—[a]	33±5
Reactor R_1 hydraulic retention time of 0.8 h				
N	4	2.48±0.37	—	41±2
C+N	5	2.49±0.37	—	—
C+N	6	2.54±0.41	0.46±0.01	44±3
Reactor R_2 Hydraulic retention time of 5.0 h				
N	4	2.43±0.40	—	42±3
C+N	5	2.72±0.40	—	—
C+N	6	3.63±0.38	0.61±0.04	56±5

[a] — = not determined.

the ammonium removal continuously decreased to zero within 50 days (Fig. 4C), while the dissolved oxygen concentration simultaneously increased (Fig. 4B). A constant COD removal rate of $0.61\,kg\,m^{-3}\,d^{-1}$ was reached within 5 days in both reactors (Figs. 4B and D) and no nitrite accumulation was observed during the entire phase III. The increase of biofilm thickness and mass after carbon addition was more pronounced in reactor R_2 than in R_1 (Table 2).

The amount of biomass found in the reactors during phase III was by a factor of 10 to 40 higher than the value expected for theoretical equilibrium between bacteria growth rate and dilution rate (Table 2). This inconsistency was obviously caused by an accumulation of mostly heterotrophic biomass in the reactors' bottom and effluent tubes increasing the actual sludge retention time considerably. In reactor R_2 this effect was supported by the lower hydraulic retention time, leading to a higher biomass accumulation than in R_1. The accumulated biomass was removed once per week and included in the samples for suspended solids quantifications and FISH analyses of suspended cells.

After the discontinuation of acetate addition (phase IV) the ammonium removal in both reactors recovered and after a period of 14 days approx. 90% of the influent ammonium load was nitrified. In reactor R_2 the dissolved oxygen concentration decreased back to the value at the beginning of phase III.

3.2. Diversity of nitrifying bacteria in the reactors

The ammonia-oxidizing cells in all biofilm samples could be labeled simultaneously with probes BET42a, Nso1225 and Nso190. No ammonia-oxidizers belonging to the *Nitrosospira*-cluster were detected. A fraction of the ammonia-oxidizing population was detectable with probe NEU, while *Nitrosococcus mobilis* was absent. The NEU-positive subpopulation of ammonia-oxidizers is most likely affiliated with the *Nitrosomonas europaea/eutropha* group [17]. Comparative sequence analyses of *amoA* clones derived from the biofilm of reactor R_0 independently confirmed the presence of two different groups of ammonia-oxidizers (Fig. 5). One *amoA* sequence cluster is closely related to *Nitrosomonas europaea*, most likely representing the NEU-positive ammonia-oxidizers, while the other *amoA* cluster is not closely related with any described ammonia-oxidizer reference strain. In both reactors the nitrite-oxidizing cells in the biofilm were affiliated with the genus *Nitrospira* during all phases of operation. Members of the genus *Nitrobacter* were only detected in the biofilm from reactor R_2 (5.0 h retention time) during operation with acetate addition (Table 3).

3.3. Diversity of heterotrophic bacteria in the reactors

The heterotrophic bacteria present in the biofilm of reactor R_2 were *Proteobacteria* of the alpha- and beta-subclasses while in reactor R_1 only beta-subclass *Proteobacteria* could be detected (Table 3). In both reactors the microbial populations in the suspended biomass during phase III were *Proteobacteria* of the alpha-, beta- and gamma-subclasses and bacteria belonging to *Cytophaga-Flavobacterium*-cluster (Table 4).

The dominating microbial populations both in biofilm and suspended biomass during combined organic carbon and ammonia oxidation belong to the beta-subclass of *Proteobacteria*. Concerning the two reactor

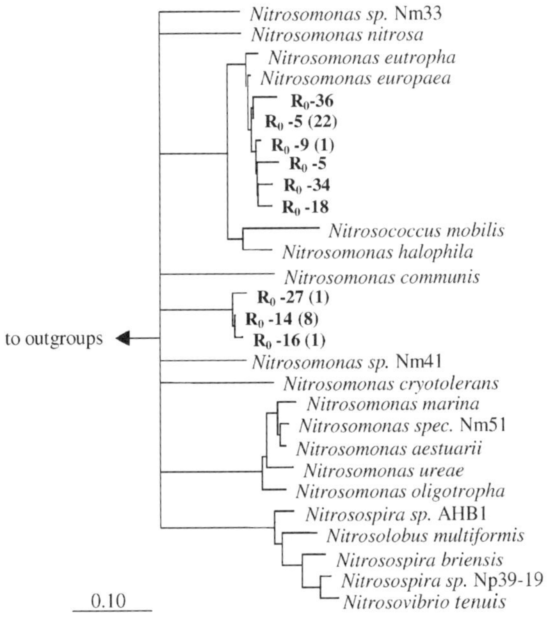

Fig. 5. Phylogenetic FITCH tree reflecting the relationships of the ammonia-oxidizers in reactor R_0 based on *amoA* sequences. The scale bar indicates the number of expected amino acid substitutions per site per unit of branch length. Numbers in brackets indicate the number of clones with almost identical sequences (>99% amino acid sequence similarity) which were retrieved from the reactor.

operation modes (N and C+N operation) the major difference is that during pure nitrification all beta-subclass *Proteobacteria* were ammonia-oxidizers, while during acetate dosing an additional presumably heterotrophic beta-subclass population developed. With respect to the different hydraulic retention times the major difference is a presumably heterotrophic alpha-subclass *Proteobacteria* population that occurred during pure nitrification only in reactor R_2 biofilm and disappeared after introduction of acetate.

3.4. Population dynamics in biofilm and suspended biomass

Values for each probe's targeted-bacteria are depicted in Table 3 for biofilm and in Table 4 for suspended biomass. The relative biovolumes of the different microbial populations (determined by FISH) were normalized by taking into account the differences in biomass (measured as dry weight) in the different reactors and samples. The FISH biovolume percentages for the sample with the highest biomass per reactor volume (sample 6) were kept unchanged and the FISH percentages of the other samples were reduced according to the biomass differences.

The fraction of the heterotrophic beta-*Proteobacteria* population (HET) present in the biofilm of reactors R_1 and R_2 can be determined as follows:

$$(\mathrm{Area}_{\mathrm{BET42a}})_{\mathrm{HET}} = \mathrm{Area}_{\mathrm{BET42a}} - \mathrm{Area}_{\mathrm{Nso1225}}. \quad (1)$$

Simultaneous hybridization of biofilm samples with probes ALF1b and Ntspa662 designed for specific detection of the alpha-subclass of *Proteobacteria* and the genus *Nitrospira*, respectively, demonstrated that members of the genus *Nitrospira* are non-specifically targeted by probe ALF1b. Consistent with this finding, a recent data base inspection demonstrated that *Nitrospira*-like 16S rRNA-sequences retrieved from wastewater treatment plants possess the full match target site of probe ALF1b. Consequently, *Nitrospiras* have to be included in the list of non-alpha-subclass *Proteobacteria* targeted by probe ALF1b [8]. The relative biovolume labeled with probe ALF1b was higher than the one labeled with probe Ntspa662 in biofilm samples taken from reactor R_2, except for the last biofilm sample during phase III (Table 3). This demonstrates that an ALF1b-positive population not related to *Nitrospira* appeared in the biofilm when the hydraulic retention time changed from 0.8 to 5.0 h, corresponding to the transition from reactor R_0 to reactor R_2 (phase Ii—N operation mode), and disappeared after the operation of reactor R_2 with acetate for 50 days (phase III—C+N operation mode).

The fraction of the alpha-*Proteobacteria* population (HET) developed in the biofilm from reactor R_2 during pure nitrification (phase II) can be estimated as follows:

$$(\mathrm{Area}_{\mathrm{ALF1b}})_{\mathrm{HET}} = \mathrm{Area}_{\mathrm{ALF1b}} - \mathrm{Area}_{\mathrm{Ntspa662}}. \quad (2)$$

The fraction of biofilm and suspended biomass bacteria identified with specific gene probes (F) in relation to all bacteria (EUB338 probe set) was calculated as listed below:

Biofilm samples

$$\mathrm{Area}_{\mathrm{ALF1b}} \approx \mathrm{Area}_{\mathrm{Ntspa662}} \Rightarrow F = \mathrm{Area}_{\mathrm{BET42a}} + \mathrm{Area}_{\mathrm{Ntspa662}}, \quad (3)$$

$$\mathrm{Area}_{\mathrm{ALF1b}} > \mathrm{Area}_{\mathrm{Ntspa662}} \Rightarrow F = \mathrm{Area}_{\mathrm{BET42a}} + \mathrm{Area}_{\mathrm{ALF1b}}. \quad (4)$$

Suspended biomass samples

$$F = \mathrm{Area}_{\mathrm{ALF1b}} + \mathrm{Area}_{\mathrm{BET42a}} + \mathrm{Area}_{\mathrm{GAM42a}} + \mathrm{Area}_{\mathrm{CF319a}}. \quad (5)$$

The population dynamics of nitrifiers and heterotrophs in the biofilm and suspended biomass from reactors R_0, R_1 and R_2 during the different phases of operation is depicted in Fig. 6. For most biofilm and suspension samples, more than 80% of the bacteria detectable by the EUB338 probe set* could simulta-

Table 3
Microbial community composition of reactors R_0, R_1 and R_2 biofilm during the different phases of operation[a]

Sample/reactor	NF	Oligonucleotide probes						F
		ALF1b	BET42a	Nso1225	NEU	Ntspa662	NIT3	
Phase I—N removal								
1 R_0	0.74	—	**39** (53±4)	**39** (53±4)	**25** (34±3)	**38** (52±4)	**0** (0)	105
2 R_0		—	—	—	(56±5)	(30±5)	0	—
3 R_0	0.74	**20** (27±3)	**41** (55±4)	**41** (55±4)	**19** (26±3)	**20** (27±3)	**0** (0)	82
Phase II—N removal								
4 R_1	0.68	**21** (31±3)	**30** (45±6)	**30** (45±6)	**16** (24±6)	**23** (33±5)	**0** (0)	78
R_2	0.67	**35** (52±3)	**21** (31±5)	**21** (31±5)	**14** (21±3)	**27** (41±2)	**0** (0)	83
Phase III—C+N removal								
5 R_1	0.68	**12** (18±3)	**26** (39±5)	**19** (27±3)	**12** (17±3)	**14** (21±2)	**0** (0)	59
R_2	0.75	**30** (40±7)	**29** (39±5)	**21** (28±7)	**19** (25±5)	**23** (30±6)	**0** (0)	80
6 R_1	0.70	**24** (34±6)	**34** (49±8)	**21** (30±4)	**5** (7±2)	**22** (32±5)	**0** (0)	81
R_2	1.00	**29** (29±6)	**60** (60±4)	**0** (0)	**0** (0)	**31** (31±6)	**3** (3±1)	94

[a] The relative biovolumes of probe-defined bacterial populations (values in brackets) and the respective normalized values in regard to the biomass content of the samples (bold values) are given in columns 3–8. No normalized values are given for sample 2 since the biomass content was not determined for this sample. The fraction of bacteria detectable with the bacterial probe set, which were identified with specific oligonucleotide probes, is given in column 9. Values listed in the table are the average percentage ±95% confidence interval. For all biofilm samples no signals were observed with probes specific for the *Cytophaga-Flavobacterium*-cluster (CF319a) and the gamma-subclass of *Proteobacteria* (GAM42a), respectively. NF = normalization factor; — = not determined.

Table 4
Microbial community composition of reactors R_1 and R_2 suspended biomass during combined nitrification and organic carbon removal (columns 3–6)[a]

Sample/reactor	NF	Oligonucleotide probes				F
		ALF1b	BET42a	GAM42a	CF319a	
Phase III—C+N removal						
5 R_1	0.18	**2.2** (12±1)	**11.5** (64±2)	**0** (0)	**0.2** (1±0)	76
R_2	0.17	**1.6** (10±1)	**14** (81±1)	**0.5** (3±0)	**0.9** (5±1)	99
6 R_1	0.18	**3.1** (17±3)	**14.6** (81±2)	**0.2** (1±0)	**0.2** (1±0)	100
R_2	0.17	**1.2** (7±1)	**13** (77±4)	**1.9** (11±1)	**1.5** (9±2)	104

[a] Column 7 displays the fraction of bacteria detectable with the bacterial probe set which were identified with specific oligonucleotide probes. Values listed in the table are the average percentage ±95% confidence interval (values in brackets). NF = normalization factor.

neously be classified by at least one of the more specific probes applied (Tables 3 and 4; Eqs. 3–5). However, this does not hold true for the biofilm and suspended biomass samples taken from reactor R_1 during the accidental overloading period with ammonium (sample 5). In this sample 41% of the biofilm bacteria and 24%

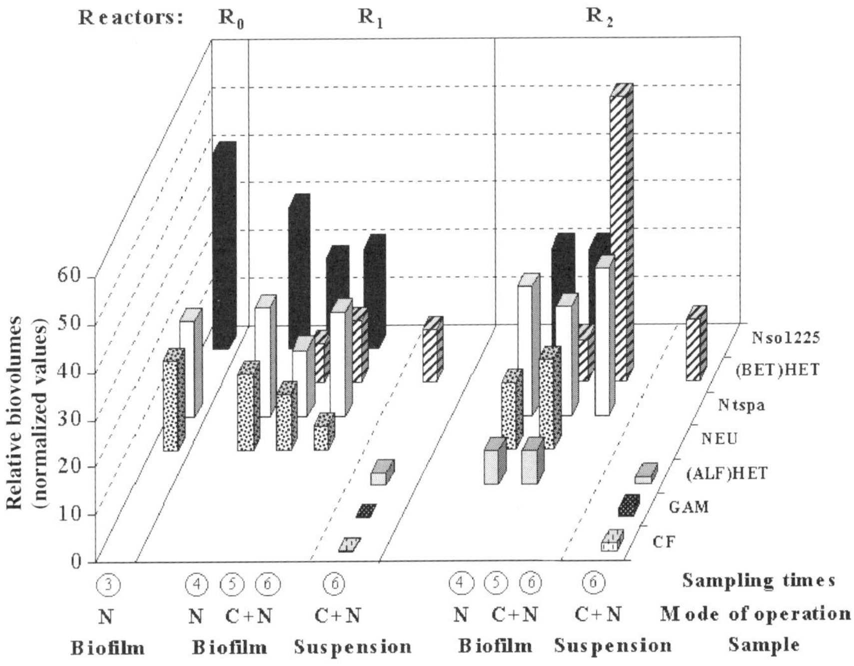

Fig. 6. Population dynamics of ammonia-oxidizers, nitrite-oxidizers and heterotrophs in the biofilm during process operation of reactors R_1 and R_2 with pure nitrification [phase I (sampling point 3) and phase II (sampling point 4)] and combined nitrification and organic carbon removal (phase III—sampling points 5 and 6). In addition, the microbial community composition of the suspended biomass is displayed for both reactors for sampling point 6.

of the suspended bacteria which were detectable by FISH could not further be classified with the more specific probes. Whether these bacteria appeared in the reactor due to the acetate addition or the elevated ammonia concentrations could not be clarified.

3.5. Population dynamics during phase I

Biofilm samples were taken before and after the storage period preceding operation of reactor R_0 (samples 1 and 2) and after 112 days of operation (sample 3), in order to evaluate the effects of the storage period on the microbial community composition.

The samples were hybridized in situ with probes Nso1225, NEU and Ntspa662, as well as the EUB338 probe set. After the storage period (sample 2) the fraction of ammonia-oxidizers was larger than during normal operation (samples 1 and 3) while the relative abundance of nitrite-oxidizers decreased (data not shown), suggesting that the nitrite-oxidizers decayed faster than the ammonia-oxidizers under the selected storage conditions. The characterization of the biofilm from reactor R_0 during operation (sample 3) showed that the ammonia-oxidizing population detected with probe Nso1225 was able to recover a comparable relative abundance as before the storage period (sample 1), while the nitrite-oxidizing population decreased considerably from 38% (sample 1) to 20% (sample 3). Furthermore, the morphology of the nitrite-oxidizing clusters changed from big clusters present before the storage period (sample 1), to small ones mixed with net-like structures afterwards (sample 3). Despite the decrease in the relative abundance of nitrite-oxidizers, no nitrite accumulation was observed during phase I. This suggests that the nitrite-oxidizers present in net-like structures in sample 3 are more active (possibly due to a better accessibility to substrates) than those occurring in big clusters in sample 1.

3.6. Population dynamics during phase II

In the transition from phases I to II the amount of biofilm particles from reactor R_0 was split up between reactors R_1 and R_2, while the NH_4^+-N load per reactor was reduced from 1.24 to 0.48–0.61 $kg\,m^{-3}\,d^{-1}$. Despite the lower amount of biofilm particles in R_1 and R_2 than in R_0, the biomass concentration per reactor volume was similar in both reactors during phases I and II (Table 2). This can be explained by a lower biofilm detachment rate in reactors R_1 and R_2 caused by a decrease in the collision frequency between biofilm particles. Consequently, the NH_4^+-N load to biomass ratio was higher in R_0 (0.46 kg kg^{-1} d^{-1}) than in R_1 (0.25 kg kg^{-1} d^{-1}) and R_2 (0.24 kg kg^{-1} d^{-1}). As a result the effluent NH_4^+-N concentration of the reactors dropped from 1.8 mg l^{-1} (R_0) to 0.6 mg l^{-1} (R_1) and 0.2 mg l^{-1} (R_2), respectively. The normalized abundance of ammonia-oxidizers in the biofilm decreased from reactor R_0 (sample 3) to reactor R_1 (sample 4) and reactor R_2 (sample 4), respectively (Fig. 6; Table 3). The NH_4^+-N concentration in reactors R_1 and R_2 was lower than the assumed saturation constant of 0.7 mg l^{-1} [18], and thus most likely limited the amount of ammonia-oxidizers in the biofilm. Unlike the ammonia-oxidizing population the relative abundance of nitrite-oxidizers increased in R_2 and remained approximately constant in R_1 during phase II (N operation). According to Eq. (2), in reactor R_2 a heterotrophic population (8%) detectable with probe ALF1b developed.

3.7. Population dynamics during phase III

The newly introduced supply of acetate during phase III induced the growth of heterotrophic microorganisms both in suspension and in the biofilm in both reactors. Biofilm samples collected 12 days after acetate addition (sample 5) showed the presence of a thin layer of heterotrophic beta-*Proteobacteria* distributed discontinuously on top of the nitrifying biofilm, accounting for 7–8% of the bacterial biofilm population in both reactors. Like during phase II, a heterotrophic population detectable with probe ALF1b (7%) was exclusively present in reactor R_2, 12 days after addition of acetate. During the first 12 days of phase III the abundance of nitrite-oxidizers decreased in the biofilm of both reactors, while the ammonia-oxidizers decreased only in R_1. The latter observation corresponds with the higher loss of ammonium removal observed in reactor R_1, (28% loss) compared to reactor R_2 (19% loss).

The ammonium removal in reactor R_2 continued to decrease during the subsequent days of acetate addition until a complete breakdown of the nitrification process happened. After 50 days of acetate addition (sample 6), no ammonia-oxidizing bacteria could be detected in reactor R_2 while the relative in situ abundance of *Nitrospira*-like nitrite-oxidizers was similar to the one during phase II (N operation). At that time reactor R_2 biofilm was dominated by heterotrophic beta-*Proteobacteria* that amounted to 60% of the total biovolume labeled with the bacterial probe set. In contrast to reactor R_2, the extended addition of acetate did not cause a complete breakdown of nitrification in reactor R_1. In this reactor, the ammonium removal efficiency stabilized at 69% and the heterotrophic beta population in the biofilm amounted to a maximum biovolume fraction of only 13% (sample 6). As in reactor R_2, the nitrite-oxidizing population in reactor R_1 did recover its initial relative abundance during prolonged acetate dosage.

4. Discussion

In this study the microbial community composition and dynamics in two nitrifying biofilm reactors (differing in hydraulic retention time) was monitored using molecular tools during a shift in process operation from pure nitrification to combined nitrification and organic carbon removal. In general, the dynamics of the microbial communities correlated well with the performance of the respective reactors. In the following sections several interesting findings are discussed in more detail.

4.1. Composition and dynamics of bacterial populations in the reactors

In both reactors at least two populations of beta-subclass ammonia-oxidizers were present. As demonstrated by oligonucleotide probing and comparative AmoA sequence analysis, one of these populations was closely related to the model organism *Nitrosomonas europaea*, while the other population surprisingly showed no close relationship with recognized ammonia-oxidizers. Nitrite oxidation was catalyzed in both reactors mainly by *Nitrospira*-like bacteria confirming the recently recognized importance of these bacteria for nitrite oxidation in several environments (e.g. [13,14,19]). In both reactors Nitrospiras occurred in previously not observed net-like structures in the biofilm. In the present work, *Nitrobacter* could only be detected in small numbers in the biofilm of reactor R_2 during simultaneous addition of acetate and ammonium (phase III) and a concurrent increase of the NO_2^--N concentration from 0.02 mg l^{-1} (phase II) to 0.39 mg l^{-1} (phase III). This result is consistent with the recently published hypothesis that Nitrospiras are k-strategists (and thus thrive at low nitrite concentrations) while *Nitrobacter* as r-strategist can compete successfully only in environments with relatively high nitrite concentrations [19]. The absence of detectable Nitrobacters in reactor R_1 during phase III might have been caused by the lower NO_2^--N accumulation (0.22 mg l^{-1}) in this reactor compared to reactor R_2. In addition, *Nitrobacter* might have benefited in phase III in R_2 from its capability to grow mixotrophically with acetate while Nitrospiras might not possess this capability [14]. On the other hand we yet have no explanation for the observed increase of *Nitrospira*-like nitrite oxidizers during phase III in both reactors despite the decreasing or even failing nitrification during this period.

Despite the lack of organic carbon dosing during phase II a presumably heterotrophic population affiliated with the alpha-subclass *Proteobacteria* was present in the biofilm of reactor R_2 (5.0 h retention time), but disappeared during reactor operation with acetate (end of phase III). Most likely this population was growing on soluble microbial products (SMP) formed by active nitrifiers in the biofilm [20] since SMP concentration should be higher in reactor R_2 than in reactor R_1 due to the difference in retention time. In phase III the supply of acetate provoked the breakdown of the nitrification process in reactor R_2, consequently SMP were no longer produced and the alpha-proteobacterial population disappeared from the biofilm.

During combined nitrification and organic carbon removal in phase III a higher group-level diversity of heterotrophic microorganisms was found in suspension than in the biofilm. This result might be explained by the fact that the heterotrophic biofilm community is exposed to a higher shear stress compared to the suspended consortia. Thus, only fast growing heterotrophs can maintain themselves in the outer layer of the biofilm, a selection pressure which might reduce diversity.

4.2. Population dynamics versus reactor performance

During phase II (pure nitrification) the biofilm was, as expected, dominated by ammonia- and nitrite-oxidizing bacteria in both reactors. After acetate addition (phase III), the formation of a thicker layer of heterotrophic bacteria in reactor R_2 (with long retention time) on the surface of the nitrifying biofilm led to increased oxygen mass transfer resistance from bulk liquid to the nitrifiers. This coincided with a drastic reduction in the ammonia-oxidizing population and a subsequent breakdown of the nitrification process. In contrast, reactor R_1 operated with short retention time displayed a smaller increase of heterotrophic biofilm bacteria on the surface which corresponded to the less pronounced reduction in nitrification performance. An alternative hypotheses for the breakdown of the nitrification in reactor R_2 during phase III would be that nitrifiers but not the heterotrophs were lost from the biofilm due to detachment and finally washed out of the reactor. This would be consistent with the absence of in situ detectable ammonia-oxidizers in sample 6 of reactor R_2. However, there are two lines of evidence that oxygen limitation and not selective washout caused the nitrification breakdown. Firstly, the even slower growing nitrite-oxidizers still were detected in significant amounts (31%) within the biofilm of reactor R_2 at the end of phase III. Secondly, the re-establishment of 90% of the nitrification capacity within 14 days after stop of acetate addition (Fig. 4; phase IV) in reactor R_2 contradicts the possibility of a previous complete depletion of ammonia-oxidizers. The mass accumulation rate for nitrifying biofilm of $0.03\,kg\,m^{-3}\,d^{-1}$, [21] excludes the chance of such a fast recovery by re-growth. Keeping in mind previously published data [2] our findings were unexpected since we assumed that reactor R_2 will, due to its longer retention time, favor suspended growth of the fast reproducing heterotrophs and thus allow for a higher biofilm-mediated nitrification during the presence of acetate.

What remains to be discussed is (i) why a thicker heterotrophic biofilm is formed on top of the nitrifying biofilm in reactor R_2 compared to reactor R_1 and (ii) why heterotrophic biofilm formation and subsequent loss of nitrifying capacity was not observed in the airlift reactor of [2] after addition of organic carbon. Heterotrophic biofilm formation in reactor R_2 could be explained by increased liquid phase viscosity in this reactor due to the presence of extracellular biopolymers which accumulated compared to reactor R_1 due to the increased liquid retention time in reactor R_2. Increased liquid viscosity will lead to a stronger air bubble coalescence, thus decreasing the volume occupied by the gas phase, reducing the driving force for the circulation and ultimately the shear stress for the biofilm ([22,23]). Consistent with this argumentation, surface protuberances were observed by us microscopically in the biofilm of R_2 (reflecting the lower shear stress) while the biofilm of reactor R_1 was characterized by a much smoother biofilm surface (indicative for a higher shear stress [23]). Differences in liquid phase viscosity or other factors influencing turbulence and thus shear stress and substrate availability (e.g. reactor geometry, friction, structure of the biofilm support particles etc.) could also be responsible for the inconsistent results between this study and the work of van Benthum and co-workers. Independent from the actual reason(s) causing these inconsistent results of both studies, it is important to note that extension of hydraulic retention time caused dramatically different effects in similar biofilm reactors.

5. Conclusions

The following main conclusions can be drawn from the present study:

1. No major effect of the hydraulic retention time on the diversity of nitrifying bacteria in the biofilm was observed. A group of ammonia-oxidizers not closely related to any described reference strain was identified in the biofilm from both reactors showing that even in systems working with defined conditions, the bacterial diversity is not completely described.
2. Combined nitrification and carbon removal under oxygen limiting conditions could be accomplished in the biofilm reactor with low hydraulic retention time but failed in the reactor with high hydraulic retention

time. This unexpected finding was caused by the formation of a thick heterotrophic layer on top of the nitrifying biofilm in the latter reactor that limited the nitrifiers' oxygen supply. Thus, extension of the hydraulic retention time is not always sufficient to improve combined nitrification and organic carbon removal in biofilm reactors.

3. Today, the battery of molecular tools allows to precisely determine ecological key parameters of complex microbial communities present in engineered systems. In addition to species richness and evenness also the in situ activity of probe identified bacteria can be analyzed [24]. Future interdisciplinary research at the interface between molecular microbial ecology and civil engineering will almost certainly allow for a detailed understanding of the links between microbial diversity, process efficiency and process stability. For example, it should be possible by the use of molecular methods to define operational parameters which selectively increase the diversity within important functional groups of bacteria (e.g. nitrifiers) and thus render the microbial community more resistant against perturbations. Armed with such knowledge innovative strategies for process control and design as well as for bioaugmentation can be developed.

Acknowledgements

This work was supported by the Portuguese Ministry of Science and Technology (PRAXIS XXI BD/15943/98), the Deutscher Akademischer Austauschdienst (A/99/06961) and the European Comission (T.M.R. program, BioToBio project). Ulrike Purkhold was supported by Sonderforschungsbereich 411 from the Deutsche Forschungsgemeinschaft (Research Center for Fundamental Studies of Aerobic Biological Wastewater Treatment; project A2).

References

[1] Tijhuis L, Rekswinkel HG, van Loosdrecht MCM, Heijnen JJ. Dynamics of population and biofilm structure in the biofilm airlift suspension reactor for carbon and nitrogen removal. Water Sci Technol 1994;29:377–84.

[2] van Benthum WAJ, van Loosdrecht MCM, Heijnen JJ. Control of heterotrophic layer formation on nitrifying biofilms in a biofilm airlift suspension reactor. Biotechnol Bioeng 1997;53(4):397–405.

[3] Ohashi A, de Silva DGV, Mobarry B, Manem J, Stahl D, Rittmann BE. Influence of the substrate C/N ratio on the structure of multi-species biofilms consisting of nitrifiers and heterotrophs. Water Sci Technol 1995;32(8):75–84.

[4] Furumai H, Rittmann BE. Evaluation of multiple-species biofilm and floc processes using a simplified aggregate model. Water Sci Technol 1994;29(10–11):439–46.

[5] Lazarova V, Manem J. An innovative process for waste water treatment: the circulating bed reactor. Water Sci Technol 1996;34(9):89–99.

[6] APHA. Standard methods for the examination of water and wastewater. 19th ed. New York: American Public Health Association, 1995.

[7] Daims H, Brühl A, Amann R, Schleifer K-H, Wagner M. The domain—specific probe EUB338 is insufficient for the detection of all *Bacteria*: development and evaluation of a more comprehensive probe set. Syst Appl Microbiol 1999;22:434–44.

[8] Manz W, Amann R, Ludwig W, Wagner M, Schleifer K-H. Phylogenetic oligonucleotide probes for the major subclasses of *Proteobacteria*: problems and solutions. Syst Appl Microbiol 1992;15:593–600.

[9] Manz W, Amann R, Ludwig W, Vancanneyt M, Schleifer K-H. Application of a suite of 16S rRNA-specific oligonucleotide probes designed to investigate bacteria of the phylum *Cytophaga-Flavobacter-Bacteroides* in the natural environment. Microbiol 1996;142:1097–106.

[10] Wagner M, Rath G, Amann R, Koops H-P, Schleifer K-H. In situ identification of ammonia-oxidizing bacteria. Syst Appl Microbiol 1995;18:251–64.

[11] Wagner, M, Rath G, Koops H-P, Flood J, Amann R. In situ analysis of nitrifying bacteria in sewage treatment plants. Water Sci Technol 1996;34(1):237–44.

[12] Mobarry BK, Wagner M, Urbain V, Rittmann BE, Stahl DA. Phylogenetic probes for analysing abundance and spatial organization of nitrifying bacteria. Appl Environ Microbiol 1996;62(6):2156–62.

[13] Juretschko S, Timmermann G, Schmid M, Schleifer K-H, Pommerening-Röser A, Koops H-P, Wagner M. Combined molecular and concentional analyses of nitrifying bacterium in activated sludge: *Nitrosococcus mobilis* and *Nitrospira*-like bacteria as dominant populations. Appl Environ Microbiol 1998;64(8):3042–51.

[14] Daims H, Nielsen P, Nielsen JL, Juretschko S, Wagner M. Novel *Nitrospira*-like bacteria as dominant nitrite-oxidizers in biofilms from wastewater treatment plants: diversity and in situ physiology. Water Sci Technol 2000;41(4–5):85–90.

[15] Schmid M, Twachtmann U, Klein M, Strous M, Juretschko S, Jetten M, Metzger J, Schleifer K-H, Wagner M. Molecular evidence of genus level diversity of bacteria capable of catalysing anaerobic ammonium oxidation. Syst Appl Microbiol 2000;23:93–106.

[16] Purkhold U, Pommerening-Röser A, Juretschko S, Schmid MC, Koops H-P, Wagner M. Phylogeny of all recognized species of ammonia-oxidizers based on comparative 16S rRNA and amoA sequence analyses: implications for molecular diversity surveys. Appl Environ Microbiol 2000;66(12):5368–82.

[17] Wagner M, Noguera DR, Juretschko S, Rath G, Koops H-P, Amann R. Combining fluorescent in situ hybridization (FISH) with cultivation and mathematical modeling to study population structure and function of ammonia-

oxidizing bacteria in activated. Water Sci Technol 1998;37(4–5):441–9.

[18] Wiesmann U. Biological nitrogen removal from wastewater. In: Fietcher A, editor. Advances in biochemical engineering biotechnology, vol. 51. Berlin Heidelberg Germany: Springer, 1994. p. 113–54.

[19] Schramm A, de Beer D, van den Heuvel JC, Ottengraf S, Amann R. Microscale distribution of populations and activities of *Nitrosospira* and *Nitrospira* spp. along a macroscale gradient in a nitrifying bioreactor: quantification by in situ hybridization and the use of microsensors. Appl Environ Microbiol 1999;65(8):3690–6.

[20] Rittmann BE, Regan JM, Stahl DA. Nitrification as a source of soluble organic substrate in biological treatment. Water Sci Technol 1994;30(6):1–8.

[21] Garrido-Fernández J, Méndez R, Lema JM, Lazarova V. The circulating floating ber reactor: effect of particle size distribution of the carrier on ammonia convertion. Conference on Biofilm Systems. New York, 1999, 17–20 October.

[22] Akita K, Yoshida F. Gas holdup and volumetric mass transfer coefficient in bubble columns. Ind Eng Chem, Process Des Dev 1993;12:76–80.

[23] Kwok WK, Picioreanu C, Ong SL, van Loosdrecht MCM, Heijnen JJ. Influence of biomass production and detachment forces on biofilm structures in a biofilm suspension airlift reactor. Biotechnol Bioeng 1998;58(4):400–7.

[24] Lee N, Nielsen PH, Andreasen KH, Juretschko S, Nielsen JL, Schleifer K-H, Wagner M. Combination of fluorescent in situ hybridization and microautoradiography—a new tool for the structure-function analyses in microbial ecology. Appl Environ Microbiol 1999; 65(3):1289–97.

PERGAMON

Water Research 36 (2002) 482–490

WATER RESEARCH

www.elsevier.com/locate/watres

Sludge population optimisation: a new dimension for the control of biological wastewater treatment systems

Zhiguo Yuan*, Linda L. Blackall

The Advanced Wastewater Management Centre, The University of Queensland, St Lucia 4072, Qld, Australia

Received 28 July 2000; received in revised form 8 January 2001; accepted 30 March 2001

Abstract

The activated sludge comprises a complex microbiological community. The structure (what types of microorganisms are present) and function (what can the organisms do and at what rates) of this community are determined by external physico-chemical features and by the influent to the sewage treatment plant. The external features we can manipulate but rarely the influent. Conventional control and operational strategies optimise activated sludge processes more as a chemical system than as a biological one. While optimising the process in a short time period, these strategies may deteriorate the long-term performance of the process due to their potentially adverse impact on the microbial properties. Through briefly reviewing the evidence available in the literature that plant design and operation affect both the structure and function of the microbial community in activated sludge, we propose to add sludge population optimisation as a new dimension to the control of biological wastewater treatment systems. We stress that optimising the microbial community structure and property should be an explicit aim for the design and operation of a treatment plant. The major limitations to sludge population optimisation revolve around inadequate microbiological data, specifically community structure, function and kinetic data. However, molecular microbiological methods that strive to provide that data are being developed rapidly. The combination of these methods with the conventional approaches for kinetic study is briefly discussed. The most pressing research questions pertaining to sludge population optimisation are outlined.

Keywords: Activated sludge; Sludge population; Optimisation; Control; Microbial community structure; Kinetics; Selection; Decay rate; Yield; Fluorescence in situ hybridisation; Batch test

1. Introduction

Considerable research has been devoted in the past decades to the optimisation and control of biological wastewater treatment processes. The designed control systems range from a single-loop controller, which focuses on an isolated process unit, to the integrated plant-wide control systems, which optimise the performance of the overall plant. A common feature of these control strategies is that the process is viewed more as a chemical one than as a biological one. The control systems are usually designed such that the pollutant (BOD, N and P) concentrations in the effluent are maintained below pre-specified levels with a minimum operational cost. The monitored variables are typically those of a chemical (e.g. NH_4^+-N, NO_3^--N, PO_4^--P and dissolved oxygen concentrations) and/or physical (e.g. flow rates) nature. Although biological variables such as NADH [1] are also sometimes used as the monitored and/or controlled variables, the impact of the control actions on the microbial community in the system and on the properties of microbes have never been explicitly taken into consideration. While the control systems may be able to 'optimise' the process in short time periods, they may deteriorate the long-term performance of the

*Corresponding author. Tel.: +61-7-3365-4374; fax: +61-7-3365-4726.

E-mail address: zhiguo@awmc.uq.edu.au (Z. Yuan).

0043-1354/02/$ - see front matter
PII: S0043-1354(01)00230-5

plants if the microbial community in the system and/or the microbial properties of sludge are adversely affected.

Indeed, mounting evidence is available indicating that the microbial properties of the system, and even the species/subspecies of microorganisms present in a system, may be affected by the operation of the system. Through briefly reviewing and discussing these evidences, we argue that sludge population optimisation should be an explicit aim to be considered in the design and operation of a treatment system. A control system should not only optimise the biological reactions in short time periods, but also, and perhaps more importantly, optimise the microbial properties of the sludge. The latter prevents shortsighted optimisation, ensuring consistent performance of the system over long terms.

The advanced molecular monitoring techniques such as whole cell or fluorescence in situ hybridisation (FISH) probing [2] are powerful tools for the study of sludge population optimisation. FISH can be used to identify and quantify, in situ, the microorganisms present in activated sludge [3–5]. The combination of these techniques with the conventional batch assays to study the microbial properties of activated sludge is discussed.

2. Sludge population optimisation: the concept

To illustrate the ideas, we take the aeration control to a nitrification process as an example:

$$NH_4^+ + O_2 \xrightarrow{X_A} NO_3^- + H_2O + 2H^+.$$

The volumetric ammonia removal rate is determined by,

$$r_{NH} = \frac{\mu_{A,max} X_A}{Y_A} \frac{S_{NH}}{K_{NH} + S_{NH}} \frac{S_O}{K_O + S_O}, \tag{1}$$

where S_{NH} and S_O are the ammonia nitrogen and oxygen concentrations, respectively; K_{NH} and K_O are the affinity constants for ammonia and oxygen, respectively; $\mu_{A,max}$ and Y_A are the maximum specific growth rate and the yield coefficient of nitrifiers, respectively; X_A is the nitrifier concentration and is determined as

$$X_A = \frac{Y_A L_N}{b_A + 1/\theta_X}, \tag{2}$$

where b_A is the decay rate of nitrifiers; θ_X is the sludge age or sludge retention time (SRT) applied to the plant; L_N is the average of r_{NH} (Eq. (1)) over a long term (a few sludge ages).

The conventional aeration control problem is often formulated as to control the aeration such that the following objective function is minimised:

$$\int_t^{t+T} (-w_r r_{NH}(\tau) + w_c c_{MV}(\tau))\, d\tau \tag{3}$$

where c_{MV} is the cost incurred by manipulating aeration; $w_r > 0$ and $w_c > 0$ are the weighting factors for the substrate removal and operational cost, respectively; t is the current time; and T is the optimisation interval, which is often chosen in the order of hours.

The above control formulation optimises the process essentially through optimising the substrate concentrations S_{NH} and S_O, assuming that the kinetic parameters $\mu_{A,max}$, Y_A, b_A, K_{NH} and K_O are not affected by the operation, which is fundamentally flawed. For example, parameter b_A is strongly influenced by the electron acceptors [6,7], which will be further discussed later.

In general, while the impact of the process operation on the microbial properties may not be seen immediately, a long-term application of a specific operational strategy may:

- apply a selection pressure which favours the growth of one group of microorganisms while disadvantaging others. A gradual change of the microbial community may thus occur in the sludge. The relevance of such a change lies in the potential to select the desirable microorganisms to improve the pollutant removal efficiency. Different species of bacteria most likely possess different in situ growth kinetics (specific growth rate, affinity constants, etc.) as well as other properties (e.g. robustness to environmental disturbance).
- result in a change to the physiological state of individual organisms, thus changing the bacterial properties even without inducing a change in community structure.

Sludge population optimisation aims at optimising the performance of a system through optimising the following properties of the sludge:

- *Sludge kinetics*: As indicated in Eq. (1), higher specific growth rates and lower affinity constants are obviously desirable for higher reaction rates. The impact of decay rates is contradictory in that a lower decay rate increases the active biomass concentration (see Eq. (2)) and hence increases the reaction rate. However, it will also lead to a higher sludge production, increasing the cost for sludge treatment and disposal. Given the fact that nitrification is typically the limiting step in a treatment process, while the heterotrophs form a large part of the excess sludge, it would be ideal to have a low autotrophic and a higher heterotrophic decay rates.
- *Yields*: Introducing Eq. (2) into Eq. (1), we see that Y_A does not affect the substrate removal rate. Given the fact that a higher yield leads to a higher sludge production, a lower yield coefficient (in particular for heterotrophs) is desirable.
- *Robustness*: Microorganisms that are more robust to environmental disturbance (both the abiotic factors

and the toxins) are desirable. A successful selection of such organisms, in particular those responsible for nitrification, will reduce the vulnerability of the plant. Another way to increase the robustness would be to have enhanced biodiversity in the microbial community such that species in the same functional group but with different optimal growth conditions are present in the system and work under different operating conditions.

- *Settleability*: Secondary settling has long been regarded as one of the bottlenecks in a treatment system. Selecting floc-forming bacteria against filamentous ones will improve the sludge settleability.

3. Plant design and operation influencing the community structure in a treatment system

3.1. Community structure in activated sludge

The microbial community of activated sludge has been quite well described in terms of both the microorganisms capable of being isolated in pure culture and the true in situ picture. The "big 4" well known divisions (from the fact that we have many pure cultures of them) in the bacterial domain, *Proteobacteria*, *Actinobacteria*, *Cytophaga-Flavobacteria*, and low G+C gram positive bacteria typically comprise the vast bulk of the sludge microbial community when FISH is used as the method of investigation. Generally, microorganisms from the β *Proteobacteria* are one specific dominant group in activated sludge [8]. However, the β *Proteobacteria* comprise many genera with many different phenotypes. Most of the nitrifiers [3], many organic carbon utilisers and at least the only known polyphosphate accumulating organism (PAO) [5] all come from the β *Proteobacteria*. Some known denitrifiers [9] come from the α *Proteobacteria*.

A specific function may be performed by a large group of bacteria. For example, 16 species from two genera belonging to the β *Proteobacteria*, namely *Nitrosomonas* (formerly *Nitrosococcus mobilis* and *Nitrosomonas*) and *Nitrosospira* (formerly *Nitrosospira*, *Nitrosovibrio* and *Nitrosolobus*) have been recognised so far to be able to catalyse ammonia oxidation [4], while another four genera, namely *Nitrobacter*, *Nitrospina*, *Nitrococcus* and *Nitrospira*, have been recognised to be able to catalyse the nitrite oxidation [4]. It has also been observed that different types of bacteria could be dominating in different treatment plants. *Nitrosomonas* is widely reported as the dominating ammonia-oxidising bacteria in many activated sludge systems [10]. However, [4] identified *Nitrosospira* as the dominating ammonia-oxidiser in a nitrifying fluidised bed reactor, while [3] reported that *Nitrosococcus mobilis* (a specific member of the genus *Nitrosomonas*) dominated an industrial wastewater treatment plant. In the aggregates taken from the inlet region of a fluidised bed reactor, [11] detected a very dense layer of ammonia oxidising cells using probe NSO190, which is a general probe for ammonia-oxidisers. However, no cells were detected with the more specific probe NSM156 (specific for *Nitrosomonas*) or NSV443 (specific for *Nitrosospira*). This could indicate that the dominating ammonia-oxidisers were neither *Nitrosomonas* nor *Nitrosospira*.

The dominance of different species is likely caused by the presence of different selection pressures in these plants. Selection takes place in a wastewater treatment plant as a result of the process conditions applied. In general, the microbial community of an activated sludge system is influenced by:

- The influent characteristics. (1) Different substrates may support the growth of different bacteria. A convincing example is the change of microbial composition in activated sludge caused by external carbon source addition to the anoxic reactor of a nitrogen removal plant. Typically, a period of one to three sludge ages is needed for the sludge to be fully adapted to methanol (a commonly used external carbon source), during which methanol degraders develop. The formation of the methanol degraders is supported by a significantly lower capacity of the methanol-adapted sludge to denitrify with acetate and other short-chain fatty acids, and a significantly higher capacity to denitrify with methanol and ethanol [12]. (2) The wastewater may contain specific compounds that may inhibit the growth of certain species. It may also contain bacteria that are enriched during the transportation of the wastewater and which continuously seed the plant. The pH and temperature of the influent may also affect the growth of microorganisms in the plant.
- The environmental conditions. The atmospheric temperature, together with the influent temperature, virtually determines the temperature in the mixed liquor. It affects the kinetics of the microorganisms.
- The design and operation of the system, which will be discussed in detail below.

3.2. Metabolic selection

Selection of microorganisms through their ability to produce energy in specific conditions has actually served as a basis for biological nutrient removal (BNR) plant designs. The periodic exposure of the microbial community to anaerobic, anoxic and aerobic conditions has enabled the selection of nitrifiers, denitrifiers and PAOs in one reactor system so that COD, N and P removal, which is rather complicated and even mutually exclusive, can be accomplished in a single sludge. The wide range of operating conditions and reactor configurations present in treatment plants also causes some other changes to the microbial community in the sludge. For example, sludge bulking has been reported to occur more often in continuously operated BNR plants than in traditional aerobic COD-only removal plant [13]. Factors such as the unaerated mass fraction, frequency of transfer of microbial community between anaerobic, anoxic and aerobic conditions and the DO, nitrate and nitrite concentrations at the anoxic–aerobic transition may all apply significant selection pressures [13]. In fact, one hypothesis for the proliferation of filaments in some BNR plants is the forced transition of heterotrophic organisms between anoxic and aerobic conditions. If denitrification is not complete under anoxic conditions, the floc-formers, which denitrify nitrate to nitrogen gas via nitrite, nitric oxide and nitrous oxide, are inhibited in their oxygen uptake under subsequent aerobic conditions by the denitrification intermediates accumulated under the preceeding anoxic conditions, in particular, nitric oxide [14]. If this is the case, the optimal nitrate concentration in the anoxic zone, which is about 1 mg NO_3-N/L, found to optimise the external carbon addition [15], is likely to be a bad set-point from the prospect of sludge population optimisation. The nitrate nitrogen concentration at the outlet of the anoxic zone should probably be controlled close to zero instead, to avoid the accumulation of intermediate denitrification products.

Likewise, the volume fractions of anaerobic, anoxic and aerobic zones and the actual nitrate and DO concentrations in each zone of the bioreactor may select certain species within each functional group of organisms (nitrifiers, denitrifiers and PAOs). In addition, the location where influent is fed also tends to affect the microbial community. [16] reported that the methanol degraders formed when methanol is added to the anoxic zone have a smaller yield, a significantly higher anoxic growth rate and a lower aerobic growth rate compared to the degraders formed when substantial methanol degradation occurs under aerobic conditions.

3.3. Kinetic selection

Growth kinetics can also be used to select for or against certain types of microorganisms. SRT and influent feeding regime play important roles in kinetic selection.

Changing SRT may cause a significant change in the microbial community in a system. By extending SRT, microorganisms with lower maximum specific growth rates such as nitrifiers may be retained in the system. These microorganisms, when having higher substrate affinity (i.e. lower K_S), may actually outcompete those with higher maximum specific growth rates but lower substrate affinities when these two microbial populations share substrates. This type of competition may in fact be responsible for the uncontrolled growth of low F/M filaments in low-loaded treatment plants [17], but this will be discussed in more detail later.

The classical example of using feeding regime to affect the microbial community is in attempts to control the growth of filamentous microorganisms during bulking and foaming episodes [18]. The steady- and transient-state growth kinetics of *Zoogloea ramigera* (a floc former) and the filamentous bulking organism Type 021N were studied in pure- and dual-culture. Kinetic data from pure cultures demonstrated that Type 021N had a higher substrate affinity, lower yield coefficient and lower excess acetate assimilation rate than *Z. ramigera*. The maximum growth rate of both organisms was moderate. In dual-culture chemostat experiments, Type 021N proliferated during low ambient acetate concentrations—i.e. conditions found in completely mixed continuous flow activated sludge plants. *Z. ramigera* proliferated during periodic high acetate concentrations simulating conditions found in intermittent operation or in plants operating with selectors. [19] then developed a mathematical model for the competitive growth of Type 021N and *Z. ramigera* with the incorporation of an aerobic selector and found successful bulking control within a defined range of aerobic selector sizes only after many sludge ages of such operation.

Incorporation of a selector is a well-known technique for sludge bulking control [17]. What happens in a selector is the so-called unbalanced growth, which has attracted considerable attention in recent years [20]. As opposed to balanced growth, during which each component in the growing biomass increases by the same factor, unbalanced growth features a different synthesis rate for some of the cellular components. During operation of a plant with a selector or as a sequencing batch reactor (SBR), the biomass is subjected to alternating periods of relatively high substrate concentrations then to low substrate concentrations. Selection is placed in favour of those microorganisms which can rapidly take up the substrates during the period of exposure to relatively high substrate concentration. The carbon substrate is typically stored as intracellular polymers like glycogen or poly-β-hydroxy alkanoates (PHAs). These storage polymers are the organic carbon the microorganisms use during the subsequent period of low ambient substrate concentration. It is often reported that sludge in plants where unbalanced growth occurs has better settleability [21], possibly because of the inability of studied filaments to rapidly take up carbon during unbalanced growth conditions in comparison to floc-formers. This is illustrated in the *Zoogloea ramigera*/Type 021N example from [19] above by looking at the transient or unbalanced growth data for these two organisms. *Z. ramigera* is able to more rapidly take-up the acetate

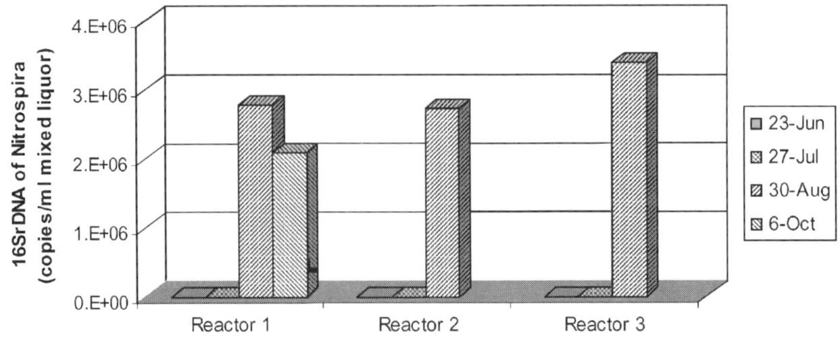

Fig. 1. Development of *Nitrospira* in three lab-scale reactors, all seeded with the same nitrifying sludge on June 23, 1999. In each measurement, total DNA was extracted from 250 μL of mixed liquor and the amount of *Nitrospira* determined by a real-time PCR assay.

under these conditions and over a relatively short time, will outcompete the Type 021N filaments.

Recently, we observed the development of *Nitrospira* populations in three lab-scale reactors. The three reactors were seeded with mixed liquor from a local treatment plant which was known to be nitrifying, and fed with the wastewater from the same plant. FISH indicated that the seeded sludge contained no detectable *Nitrospira*. However, in all three reactors, a gradual development of *Nitrospira* was observed according to a *Nitrospira*-specific real-time polymerase chain reaction (PCR) assay (Fig. 1). Given the fact that the lab-scale reactors treated the same wastewater as the full-scale plant, we hypothesise that it was the operational strategies, and likely the feeding regime, that selected *Nitrospira*. The lab-scale reactors were run as SBRs, while the full-scale plant was a continuous-flow pre-denitrification system with a completely mixed anoxic reactor and a plug-flow like aerobic reactor.

3.4. Physical selection

Certain groups of microorganisms may also be enriched in certain sections of a system through physical measures. An example for this is a process that uses biomass-growth-supporting media in bioreactors (see e.g. [22]), where the microorganisms grow as a 'biofilm'. Autotrophs and heterotrophs are physically separated in such a system. Autotrophs are selected against in the anoxic zone due to the absence of oxygen. Similarly, heterotrophs are selected against in the aerobic zone because of the lack of COD in this zone. As a result, both nitrifiers and denitrifiers are only present in the zones of the reactor where they are required, leading to a situation that the full nitrification and denitrification capacities of the plant can be achieved simultaneously. Results have shown that the maximum nitrification and denitrification rates in such a plant are significantly higher than in a conventional single-sludge system receiving the same nitrogen load [23].

The non-uniform distribution of biomass is a common feature of biofilm processes. Schramm et al. [11] compared the community structure of aggregates taken from the inlet and outlet of a lab-scale fluidised bed nitrifying reactor using FISH. In the inlet aggregates, a higher abundance and activity of ammonia oxidising bacteria relative to nitrite oxidising bacteria were observed. In the outlet aggregate, the opposite was observed. The capability of biofilm processes in separating biomass needs to be further exploited for more efficient BNR systems.

Another example for physical selection is the enrichment of active biomass through selective wastage [7]. The surplus sludge is removed from the system in such a way that the wastage rate of active biomass is lower than that of inert solids.

4. Plant design and operation influencing bacterial properties

In addition to operational strategies influencing the microbial population structure, which in turn influence the sludge properties, there is strong evidence supporting the notion that the same organisms may have different kinetics and/or stoichiometry when different operational conditions are provided. Most of the studies tend to indicate that the types of electron acceptor have the most significant influence.

4.1. Decay rate

Both the heterotrophic and autotrophic decay rates have been found electron-acceptor dependent. Some comparisons made in literature are shown in Table 1.

Table 1 indicates that:

- Both heterotrophic and autotrophic decay rates are significantly lower under anoxic condition than under

Table 1
Comparisons of bacterial decay rates under aerobic and anoxic conditions

Sources	Heterotrophs (b_H)		Autotrophs (b_A)		
	Aerobic	Anoxic	Aerobic	Anoxic	Anaerobic
[24]	0.111	0.057			
[25]			0.2	0.1	
[26]			0.46	0.09	
[27]	0.093	0.047			
[6]	0.5	0.1	0.21	0.10	0.03
[7]			0.35 when DO = 6 mg/L 0.16 when DO = 2 mg/L 0.08 when DO = 0–1.5 mg/L[a] 0.03 when DO = 0–1.5 mg/L[b]		

[a] Aeration was turned on for 15 min in every 30 min.
[b] Aeration was turned on for 15 min in every 45 min.

Table 2
Comparison of heterotrophic yield coefficients under aerobic and anoxic conditions

Sources	Substrates	$Y_{H,aerobic}$	$Y_{H,anoxic}$	$Y_{H,anoxic}/Y_{H,aerobic}$
[24]	Bacto-peptone (nutrient broth)	0.503 VSS/COD	0.272 VSS/COD	0.54
[30]	Citrate, acetate, glucose, bacto-peptone, yeast extract	0.637–0.669 COD/COD	0.192–0.512 COD/COD	0.62 (average value)
[31]	Acetic acid/starch, glucose, acetate	0.54–0.67 COD/COD	0.44–0.57 COD/COD	0.66–0.85

aerobic condition. It also appears that the decay rates are even smaller under anaerobic condition.

- The data reported in [7] suggest that the decay rate could be a continuous function of DO concentration.

The above observations are yet to be explained. Most of the decay experiments reported involved running a batch reactor for a few days, during which the bacteria were kept in starvation. A change in the microbial structure during this period was very unlikely, although not confirmed. The measured decay rates were, therefore, the different behaviours of the same microbial populations under different conditions.

Griffiths [28] suggests that the lower decay rate under anoxic conditions may be caused by the inability of protozoa to predate under such conditions. Protozoa predation has been believed by many researchers [29] to be one of the most significant processes involved in biomass decay.

4.2. Heterotrophic yield

Electron acceptors influencing heterotrophic yield has also been reported (Table 2). The heterotrophic yield coefficients observed are significantly lower for the anoxic growth than for the aerobic growth.

Smaller sludge yield in a sequential aerobic/anaerobic plant has been obtained by researchers in an effort to minimise sludge production using uncoupled anabolism and catabolism [32]. An anaerobic reactor is placed in the sludge recycling line where the recycled sludge is treated anaerobically before being fed to the aerobic reactor. A considerable part of the energy formed under oxic conditions, which would otherwise be available for biomass growth, is used for biomass maintenance functions under anaerobic conditions. Anabolic processes are thus hampered in the oxic reactor, resulting in reduction of excess sludge production. This configuration has also resulted in the selection of PAOs, which comprised of 50–60% of the whole bacterial populations.

5. Use of molecular methods to study sludge population optimisation

To enable systematic and effective sludge population optimisation, the following fundamental questions remain to be answered:

- How and why does a particular design and/or operational strategy affects the microbial community structure in an activated sludge?

- How and why are the properties of microorganisms influenced by operational conditions?
- Which microorganisms are superior than others within the same functional group (e.g. *Nitrosomonas* vs. *Nitrosospira*) in terms of kinetics, stoichiometry, floc forming capability and robustness against disturbance, and should hence be selected for?

Modern molecular monitoring techniques such as FISH are central to answering the above questions. These methods not only enable one to identify the microorganisms present in the sludge but also provide a potential means for the quantitation of these organisms and possibly, in some cases, indications of their activity. The latter, combined with the traditional techniques, enables detailed studies of the sludge kinetics, which has hardly been possible before.

5.1. Study of activated sludge kinetics by combining modern molecular monitoring techniques with conventional methods

Rate methods have been predominantly used to study the kinetics of microorganisms. Typically a batch experiment is carried out, during which the oxygen uptake rate (OUR) or substrate uptake rate (SUR) or production rate (SPR), e.g. the ammonia removal rate or nitrate production rate in a nitrifying experiment, is measured. From the OUR, SUR or SPR profiles, one works out the affinity constants as well as a variable indicating the maximum capacity of the corresponding functional group of microorganisms (e.g. nitrifiers). The latter can be, taking nitrifiers for example, the maximum exogenous oxygen uptake rate of nitrifiers $OUR_{max} = \mu_{A,max}X_A(4.57 - Y_A)/Y_A$ or the maximum ammonia uptake rate $AUR_{max} = \mu_{A,max}X_A/Y_A$ (symbols have been explained in a previous section).

One fundamental difficulty with the rate approaches is that, still taking nitrifiers as an example, parameter $\mu_{A,max}$ and biomass concentration X_A are always coupled together (as shown by OUR_{max} and AUR_{max} above). The combination is further complicated by the presence of a yield coefficient. As X_A is traditionally an unmeasurable quantity, X_A had to be estimated either explicitly [25] or implicitly [33] using Eq. (2) or its dynamic versions. The estimation requires the knowledge of b_A, the in situ autotrophic decay rate. It has been shown that the estimate of $\mu_{A,max}$ is highly sensitive to the value of b_A [33].

Determination of parameter b_A typically involves running a batch reactor with starved sludge for a period of time like one to two weeks and measuring OUR_{max} or AUR_{max} at regular intervals (e.g. daily) through adding ammonia pulses. The decay rate is then obtained by fitting the first-order decay model to the OUR_{max} or AUR_{max} data. The underlying assumption behind the calculation is that both $\mu_{A,max}$ and Y_A remain constant during the batch assay, whose substantiation requires the unavailable knowledge of $\mu_{A,max}$ and Y_A.

Quantification of biomass allows decoupling the maximum specific growth rate and the biomass concentration. This facet of wastewater treatment is currently in a development phase. However, some techniques are already available and many others are emerging.

From the simple perspective of how many microorganisms are present in a sample, direct counting of the cells following staining with DNA intercalating dyes like 3,6-tetramethyl diaminoacridine (acridine orange) [34] and 4,6-diamidino-2-phenylindole (DAPI) [35] can be employed. This, however, does not distinguish between living and dead cells as even dead cells can still harbour DNA. One of the most well used methods for the quantitation of specific microorganisms in complex microbial ecosystems is the "most probable number" (MPN). This employs a statistical approach in which successive dilutions are performed to reach an extinction point. Replicates of each dilution are prepared and a statistical table based upon a Poisson distribution is used to determine the MPN of viable organisms in the sample. MPNs have been widely used to determine numbers of nitrifiers for example.

Immunological methods employing tagged poly- or mono-clonal antibodies are also used for determining the presence and quantification of specific organisms in sludge. This procedure has been used for nitrifiers [36]. A wide range of nucleic acid based techniques is used for microorganism quantitation. The methods fall into ex situ approaches like slot or dot blot hybridisations and Southern (DNA) or Northern (RNA) hybridisations or in situ hybridisations. The reporter on the gene probe employed to indicate when DNA hybridisation has occurred can be radioactive (e.g. ^{32}P labelled nucleotides) or non-radioactive (e.g. fluorescent or colourimetric). Densitometry is typically employed as the analytical method in ex situ approaches, while quantification involves comparisons of amount of bound probe from specific probings (e.g. with division or genus level probes) with those from broad specificity probing (e.g. bacterial level probes considered to be 100%). Image analysis of fluorescence micrographs from procedures like simple AO or DAPI staining [34] or FISH is being increasingly employed as a quantitative method.

From FISH, quantification of microorganisms at genus, sub-division and division level can be determined making this a powerful method for rapid understanding of the true microbial community structure. This latter approach can be linked with aspects of the phenotype of the specific microorganisms as has been demonstrated by a combination of FISH and in situ microautoradiography [37] and by a combination of FISH with stains for cellular processes like polyphosphate accumulation

[5]. Additionally, knowledge of the kinetics of in situ processes can be obtained by carrying out microsensor experiments. Data from microsensor experiments can be linked with information from the above-mentioned microscopic methods. One can see that with a combination of all these in situ methods, a deep understanding of the cellular processes at a micro-level is possible.

However, many challenging questions are yet to be answered with regard to the quantitation of microorganisms. Firstly, what is being quantitated needs to be clarified. Is it the number of microorganisms, or their biovolume or their biomass that is required? Secondly, there is difficulty in distinguishing between active, inactive or dead microorganisms in situ. Lissamine green, a selective stain for the cytoplasm of degenerating and degenerated cells, could be used simultaneously with FISH to investigate the viability status of different types of microorganisms. Furthermore, how are the data presented—i.e. do we need to know the number/amount of specific microorganisms in relation to total bacteria or total biomass which will include Archaea and Eucarya?

6. Conclusions

Sludge population optimisation is both important and feasible. Fundamental studies to understand how certain microorganisms are selected for or against and how bacterial properties are influenced by particular plant designs and operations need to be carried out in a systematic way. Modern molecular monitoring techniques are important tools for these studies. In particular, the quantitation of microorganisms using these techniques will break new ground for studying the kinetics of microorganisms.

Closer cooperation between engineers and microbiologists is crucial to this success. The most rapid fundamental advances will come from the incorporation of detailed micro-scale data into current mathematical models such that these models more closely represent the sludge processes, allowing model-based sludge population optimisation. While definitive methods to determine both microbial community structure and function are now available and data are currently being generated by some research groups with them, a great deal of effort from both microbiologists and engineers is still required for the practical application of these methods to the quantitative study of activated sludge.

Acknowledgements

We thank Mr. Steve Pickering and Ms. Sandra Hall for the experimental results described in Fig. 1. The first author thanks CRC for Waste Management and Pollution Control Ltd., Australia, for the financial support provided.

References

[1] Armiger WB, Lee GJ, Schwegler Jr BR, Mah TJ. Control of BNR processes by regulation of the VFA/M ratio. Water Sci Technol 1993;28(11–12):303–14.

[2] Amann RI, Ludwig W, Schleifer KH. Phylogenetic identification and in situ detection of individual microbial cells without cultivation. Microbiol Rev 1995;59:143–69.

[3] Juretschko S, Timmermann G, Schmid M, Schleifer K-H, Pommerening-Röser A, Koops H, Wagner M. Combined molecular and conventional analyses of nitrifying bacterium diversity in activated sludge: *Nitrococcus mobilis* and *Nitrospira*-like bacteria as dominant populations. Appl Environ Microbiol 1998;64:3042–51.

[4] Schramm A, De Beer D, Wagner M, Amann R. Identification and activities in situ of *Nitrosopira* and *Nitrospira* spp. as dominant populations in a nitrifying fluidized bed reactor. Appl Environ Microbiol 1998;64(9):3480–5.

[5] Crocetti GR, Hugenholtz P, Bond PL, Schuler A, Keller J, Jenkins D, Blackall LL. Identification of polyphosphate-accumulating organisms and design of 16S rRNA-directed probes for their detection and quantitation. Appl Environ Microbiol 2000;66:1175–82.

[6] Siegrist I, Brunner G, Koch G, Phan LC, Le VC. Reduction of biomass decay rate under anoxic and anaerobic conditions. Water Sci Technol 1999;39(1):129–37.

[7] Yuan Z, Bogaert H, Verstraete W. Reducing the size of a nitrifying activated sludge wastewater treatment plant by shortening the retention time of inert solids. Water Res 2000;34(2):539–49.

[8] Kämpfer P, Erhart R, Beimfohr C, Bohringer J, Wagner M, Amann R. Characterization of bacterial communities from activated sludge—culture-dependent numerical identification versus in situ identification using group- and genus-specific rRNA-targeted oligonucleotide probes. Microb Ecol 1996;32:101–21.

[9] Layton AC, Karanth PN, Lajoie CA, Meyers AJ, Gregory IR, Stapleton RD, Taylor DE, Sayler GS. Quantification of Hyphomicrobium populations in activated sludge from an industrial wastewater treatment system as determined by 16S rRNA analysis. Appl Environ Microbiol 2000;66:1167–74.

[10] Wagner M, Rath G, Amann R, Koops HP, Schleifer KH. In situ identification of ammonia-oxidizing bacteria. Syst Appl Microbiol 1995;18:251–64.

[11] Schramm A, De Beer D, van den Heuvel H, Ottengraf S, Amann R. In situ structure/function studies in wastewater treatment systems. Water Sci Technol 1998;37(4–5):413–6.

[12] Hallin S, Pell M. Metabolic properties of denitrifying bacteria adapting to methanol and ethnol in activated sludge. Water Res 1998;32(1):13–8.

[13] Ekama GA, Wenzel MC. Difficulties and development in biological nutrient removal technology and modelling. Water Sci Technol 1999;39(6):1–11.

[14] Casey TG, Wenzel MC, Ekama GA, Loewenthal RE, Marais GvR. An hypothesis for the cause and control of anoxic–aerobic (AA) filament bulking in nutrient removal activated sludge systems. Water Sci Technol 1994;29(7):203–12.

[15] Yuan Z, Bogaert H, Vanrolleghem P, Thoeye C, Vansteenkiste G, Verstraete W. Control of external carbon addition to predenitrifying systems. J Environ Eng 1997;123(11):1080–6.

[16] Purtschert I, Gujer W. Population dynamics by methanol addition in denitrifying waste water treatment plants. Water Sci Technol 1999;39(1):43–50.

[17] Chudoba J, Grau P, Ottova V. Control of activated sludge filamentous bulking-II, selection of microorganisms by means of a selector. Water Res 1973;7:1389–406.

[18] van Niekerk AM. Competitive growth of flocculent and filamentous microorganisms in activated sludge systems. Ph.D., University of California, Berkeley, USA, 1985.

[19] van Niekerk AM, Jenkins D, Richard MG. The competitive growth of *Zoogloea ramigera* and Type O21N in activated sludge and pure culture—a model for low F : M bulking. J Water Pollut Control Fed 1987;59:262–8.

[20] Majone M, Dircks K, Beun JJ. Aerobic storage under dynamic conditions in activated sludge processes. The State of the Art. Water Sci Technol 1999;39(2):61–73.

[21] Chiesa SC, Irvine RL, Manning Jr JF. Feast/famine growth environments, activated sludge population selection. Biotechnol Bioeng 1985;27:562–9.

[22] Welander U, Henrysson T, Welander T. Nitrification of landfill leachate using suspended-carrier biofilm technology. Water Res 1997;31(9):2351–5.

[23] Deguchi H, Kashiwaya M. Study on nitrified liquor recycling process operations using polyurethane foam sponge cubes as a biomass support medium. Water Sci Technol 1994;30(6):143–9.

[24] McClintock SA, Sherrard JH, Novak JT, Randall CW. Nitrate versus oxygen respiration in activated sludge process. J Water Pollut Control Fed 1988;60:342–50.

[25] Nowak O, Schweighofer P, Svardal K. Nitrification inhibition—a method for the estimation of actual maximum autotrophic growth rates in activated sludge systems. Water Sci Technol 1994;30(6):9–19.

[26] Leenen E, Boogert A, van Lammeren A, Tramper J, Wijffels R. Dynamics of artificially immobilized *Nitrosomonas europaea*: effect of biomass decay. Biotechnol Bioeng 1997;55:630–41.

[27] Avcioglu E, Orhon D, Sözen S. A new method for the assessment of heterotrophic endogenous respiration rate under aerobic and anoxic conditions. Water Sci Technol 1998;38(8–9):95–103.

[28] Griffiths P. The rate of death and predation in activated sludge systems incorporating anoxic zones. Proceedings of BNR3, Brisbane, 1997.

[29] Van Loosdrecht MCM, Henze M. Maintenance, endogeneous respiration, lysis, decay and predation. Water Sci Technol 1999;39(1):107–17.

[30] Copp JB, Dold PL. Comparing sludge production under aerobic and anoxic conditions. Water Sci Technol 1998;38(1):285–94.

[31] Spérandio M, Urbain V, Audic JM, Paul E. Use of carbon dioxide evolution rate for determining heterotrophic yield and characterising denitrifying biomass. Water Sci Technol 1999;39(1):139–46.

[32] Chudoba P, Morel A, Capdeville B. The case of both energetic uncoupling and metabolic selection of microorganisms in the OSA activated sludge system. Environ Technol 1992;13:761–70.

[33] Yuan Z, Bogaert H, Devisscher M, Verstraete W. On-line estimation of the maximum specific growth rate of nitrifiers in an activated sludge system. Biotechnol Bioeng 1999;65:265–73.

[34] Byrd JJ, Colwell RR. Microscopy applications for analysis of environmental samples. In: Levin MA, Seidler RJ, Rogul M, editors. Microbial ecology: principles, methods, and applications. New York: McGraw-Hill, Inc., 1992. pp. 93–112.

[35] Porter KG, Feig YS. The use of DAPI for identifying and counting aquatic microflora. Limnol Oceanogr 1980;25:943–8.

[36] Sandén B, Grunditz C, Hansson Y, Dalhammar G. Quantification and characterisation of *Nitrosomonas* and *Nitrobacter* using monoclonal antibodies. Water Sci Technol 1994;29–7:1–6.

[37] Lee N, Nielsen PH, Andreasen KH, Juretschko S, Nielsen JL, Schleifer KH, Wagner M. Combination of fluorescent in situ hybridization and microautoradiography—a new tool for structure-function analyses in microbial ecology. Appl Environ Microbiol 1999;65:1289–97.

PERGAMON

Water Research 36 (2002) 491–500

WATER RESEARCH

www.elsevier.com/locate/watres

Population changes in a biofilm reactor for phosphorus removal as evidenced by the use of FISH

Christina M. Falkentoft[a,b], Elisabeth Müller[b], Patrik Arnz[b], Poul Harremoës[a], Hans Mosbæk[a], Peter A. Wilderer[b], Stefan Wuertz[b,*]

[a] *Department of Environmental Science and Engineering, Technical University of Denmark, Bygningstorvet, Building 115, DK-2800 Lyngby, Denmark*

[b] *Institute of Water Quality Control and Waste Management, Technical University of Munich, Am Coulombwall, D-85748 Garching, Germany*

Received 21 June 2000; received in revised form 9 December 2000; accepted 31 January 2001

Abstract

Induction of denitrification was investigated for a lab-scale phosphate removing biofilm reactor where oxygen was replaced with nitrate as the electron acceptor. Acetate was used as the carbon source. The original biofilm (acclimatised with oxygen) was taken from a well-established large-scale reactor. During the first run, a decrease in the denitrifying bio-P activity was observed after 1 month following a change in the anaerobic phase length. This was initially interpreted as a shift in the microbial population caused by the changed operation. In the second run, biomass samples were regularly collected and analysed by fluorescent in situ hybridisation (FISH) and confocal laser scanning microscopy (CLSM). Concurrently, samples were taken from the original reactor with oxygen as electron acceptor in order to investigate natural microbial fluctuations. A similar decrease in the activity as in the first run was seen after one month, although the phase lengths had not been varied. Hence, the decrease after 1 month in the first and second run should be seen as a start-up phenomenon. FISH could detect a noticeable shift in the microbial population mainly within the first 2 weeks of operation. Almost all bacteria belonging to the alpha subclass disappeared and characteristic clusters of the beta and gamma subclasses were lost. Small clusters of gram-positive bacteria with a high DNA G+C content (GPBHGC) were gradually replaced by filamentous GPBHGC. Most of the bacteria in the denitrifying, phosphate removing biofilm belonged to the beta subclass of *Proteobacteria*. The applied set of gene probes had been selected based on existing literature on biological phosphate removing organisms and included a recently published probe for a *Rhodocyclus*-like clone. However, none of the specific probes hybridised to the dominant bacterial groups in the reactors investigated. No noticeable changes were detected in the aerobic bench-scale reactor during this period, indicating that the observed changes in the lab-scale reactor were caused by the changed environment. © 2002 Elsevier Science Ltd. All rights reserved.

Keywords: Denitrification; Anoxic; Phosphorus removal; Biofilm; FISH; CLSM

1. Introduction

Removal of phosphorus from wastewater was introduced in Scandinavia in the late 1960s. In the 1970s phosphorus removal was incorporated in the wastewater treatment strategy of several countries, especially countries with many inland lakes including Sweden, Norway, Finland, Canada, USA and Switzerland. Originally phosphorus was removed chemically by precipitation and this is still the dominant removal technology; yet the use of enhanced biological phosphorus removal (EBPR) by activated sludge has increased significantly during the last decade. Biological

*Corresponding author. Tel.: +49-89-289-13708; fax: +49-89-289-13718.

E-mail address: stefan.wuertz@bv.tu-muenchen.de (S. Wuertz).

PII: S0043-1354(01)00231-7

treatment has the advantage of lower sludge production, no costs of chemicals and a more ecological image. Incorporation of EBPR in biofilter plants is still at an experimental stage. The reason for this is mainly the complication caused by the diffusion aspect [1].

The design criteria for bio-P removal plants are primarily based on empirical guidelines, and despite significant efforts it has not yet been possible to definitively identify the bacterial group(s) responsible for the biological phosphate removal process. Due to the inherent relationship between habitat and microbial community, it is necessary to combine investigations of operating conditions with analysis of the microbial population.

This study investigated induction of denitrifying activity in a lab-scale phosphate removing biofilm. Redox conditions were alternated between anaerobic and anoxic phases without any aerobic phase. The inoculum originated from an EBPR bench-scale reactor operated with oxygen as the electron acceptor. Biomass samples were regularly collected and investigated with FISH to track microbial population shifts. For comparison, samples were also taken from the aerobic bench-scale reactor as a test of natural fluctuations in a reactor not subjected to changing operating conditions.

2. Theory

2.1. Biological phosphorus removal

Phosphate accumulating organisms (PAO's) are able to take up increased amounts of phosphorus compared to the amount required for normal metabolism. The process, called enhanced biological phosphorus removal (EBPR), occurs if bacteria are challenged with alternating anaerobic (i.e. no oxygen or nitrate) and either anoxic (i.e. no oxygen) or aerobic conditions. Details of their metabolism are still not completely known [2]. Some PAOs, but apparently not all, can denitrify [3].

2.2. Microbiology of phosphate removing bacteria

Mino et al. [2] summarised the conclusions that have been made so far regarding the microbiology and biochemistry of the biological phosphate removal process. *Acinetobacter* spp. were for many years considered important bio-P organisms, but have recently been shown to constitute only a minor fraction of the bacterial phosphorus removing population in activated sludge [4–6]. The reason that *Acinetobacter* spp. were falsely identified as bio-P organisms was due to the bias introduced by the traditional culture-dependent methods used to analyse microbial communities [4]. Only by the introduction of innovative methods has it become possible to detect species present in situ that are not culturable in the laboratory. The recent findings have been that the beta subclass of *Proteobacteria* dominates most municipal wastewater sludges, both with and without EBPR [5–7]. Other major groups are the alpha subclass of *Proteobacteria*, the *Planctomycetales* and the *Flexibacter-Cytophaga-Bacteroides* group [5]. Members of the class *Actinobacteria* (Gram-positive bacteria with a high DNA G+C content, GPBHGC) were found to be the second most frequent group after the beta subclass of *Proteobacteria* [7] In EBPR sludge GPBHGC [4,8], the alpha subclass of *Proteobacteria* [8] and the *Rhodocyclus* group within the beta subclass of *Proteobacteria* [5] were the most abundant groups. Melasniemi et al. [9] reported *Micrococcus*, *Staphylococcus* and *Acidovorax* and also bacteria related to actinomycetes to be common bacterial genera in EBPR sludge. However, Hiraishi et al. [7] concluded based upon quinone profiling that bacterial communities were more influenced by wastewater characteristics than by plant operational parameters. They found larger differences in the populations for different wastewater sludges than when comparing EBPR to standard processes. This strongly argues for specifying the feed and operating conditions used for any investigation of a microbial population. Hesselmann et al. [10], worked with a sequencing batch lab-scale reactor over a 3-yr period and obtained a highly enriched culture with good phosphate removal. Bacteria related to the *Rhodocyclus* group were shown to make up 81% of the population. The set-up was operated with an aerobic phase (instead of an anoxic as applied in this study). *Rhodocyclus*-like organisms have subsequently been found in several laboratory EBPR sludges from different continents [11].

3. Materials and methods

3.1. Denitrifying lab-scale reactor

A continuous lab-scale biofilm reactor was alternated between anaerobic (302 ppm acetate-COD) and anoxic (53 ppm nitrate-N) conditions [1]. Synthetic wastewater was used. The water volume of the system was 0.27 L, and the volume of biofilm carrier particles was 0.32 L. The inlet flow was $1.065\,\mathrm{L\,h^{-1}}$, and high recirculation kept the system close to ideal mix. pH was controlled at 7 ± 0.1. The feed to the reactor was added from three different tanks. A solution with micro-nutrients and buffer was continuously added and passed a deoxygenator system to assure complete oxygen-removal [12]. Alternating conditions were obtained via a computer-controlled three-way valve allowing the addition of a concentrated acetate solution during the anaerobic phases versus the addition of a concentrated potassium nitrate solution during the anoxic phases. The phase-specific solutions were flushed with nitrogen gas upon preparation and supplied with nitrogen-filled bags in the set-up to assure oxygen-free conditions. During the shift from one phase to another, nitrate and acetate were simultaneously present until the substrate from the previous phase had been completely flushed from the system. The hydraulic residence time was 25 min. The denitrifying reactor was inoculated with biofilm-coated carrier material originating from either a pilot-scale ($17\,m^3$) sequencing batch biofilm reactor in Ingolstadt, Germany, described by Arnz et al. [13], or from the aerobic bench-scale reactor described below.

3.2. Aerobic bench-scale set-up

A sequencing batch biofilm reactor was operated for combined COD removal, nitrification and phosphate removal. The carrier was Biolith, which is expanded sintered clay-balls, 4–8 mm in diameter and with a specific surface area of $500\,m^2\,m^{-3}$. The biofilter volume was 20 L. The cycle consisted of 20 min fill, 160 min anaerobic phase, 260 min aerobic phase and 40 min draw. Presettled municipal wastewater was used (~200 ppm COD, 5–10 ppm P).

3.3. Sampling and cell fixation

Samples were collected once every week from the reactors. The biomass that detached during backwashing after an aerobic or anoxic phase was used for this purpose. Fixation was done with ethanol or paraformaldehyde (PFA) according to the protocols described by Amann [14]. The fixed samples were stored at −20°C.

3.4. Phosphate measurements

Phosphate was measured on-line in the anoxic lab-scale reactor according to Standard Methods, ASTM D 515-68 non referee method B, and in the aerobic bench-scale reactor with a P analyser, Phosphax Inter (Dr. Lange, Düsseldorf, Germany). Standard test kits for analysis of nitrogen compounds and COD in grab samples were also from Dr. Lange (type LCK, digital photometer ISIS 6000).

3.5. In situ hybridisation and oligonucleotide probes

Fixed biofilm samples were immobilised on glass slides by air drying and dehydrated for 3 min in 50, 80 and 100% (v/v) ethanol, respectively. After the dehydration step, the ethanol-fixed samples were treated with lysozyme enzyme ($100{,}000\,U\,mg^{-1}$). These pre-treated samples were subjected to probes detecting the gram-positive bacteria with high G+C DNA content (HGC69a, lysozyme: $20\,g\,l^{-1}$, 15 min), the nocardioform actinomycetes (MNP1, lysozyme: $10\,g\,l^{-1}$, 20 min), and *Microlunatus phosphovorus* (MP2, lysozyme: $20\,g\,l^{-1}$, 30 min). For probes detecting gram-negative bacteria, PFA-fixed samples were used. The hybridisation procedure was carried out as described by Amann [14]. The stringency in the hybridisation buffer and washing buffer was probe dependent and was adjusted by changing the formamide or NaCl concentration (Table 1). The ethanol-fixed samples frequently detached during the washing procedure; therefore, a modified washing step was used. Warm (48°C) washing buffer was gently added using a pipette to cover the sample on the slide surface, and the slide was then incubated in a moisture chamber for 15 min at 48°C. Then the slide was rinsed with washing buffer; new washing buffer was added to the slide followed by another 15 min of incubation. This step was repeated twice. The rRNA-targeted probes used are listed in Table 1. The probes were purchased from MWG Biotech (Ebersberg, Germany) and labelled with the sulfoindocyanine dyes Cy3 or Cy5. Due to only a single mismatch between the BET42a and GAM42a probes, unlabelled probe (unlabelled GAM42a to labelled

Table 1
Oligonucleotide probes used for in situ hybridisation

Gene probe	Specificity	Formamide (%)	NaCl (mM)	Reference
EUB338	*Bacteria*	0–50	10–900	[17]
Alf1b	Alpha subclass of the *Proteobacteria*	20	225	[18]
Bet42a	Beta subclass of the *Proteobacteria*	35	80	[18]
GAM42a	Gamma subclass of the *Proteobacteria*	35	80	[18]
CF319	*Cytophaga-Flavobacteria* group	35	80	[4]
RHC438	*Rhodocyclus*-like cluster	30	100	[10]
RHX851	*Rhodocyclus*-like clone	30	100	[10]
ACA23a	*Acinetobacter* spp.	35	80	[18]
HGC69a	Gram-positive bacteria with high DNA G+C content (GPBHGC)	20	225	[19]
MNP1	Nocardioforme actinomycetes	50	10	[20]
MP2	*Microlunatus phosphovorus*	10	490	[8]

BET42a and vice versa) was added to prevent binding to non-target cells.

3.6. Staining with fluorescent dyes

For visualisation of the biofilm thickness, samples were stained with 0.01% Flourescein 5-isothiocyanate (FITC) for 15 min followed by three washing steps (5 min each) in phosphate buffer solution (PBS). FITC binds to amino groups, whereby both cells and extracellular polymeric substances (EPS) are stained.

3.7. Confocal laser scanning microscopy (CLSM)

All confocal images were recorded using a 410 CLSM (Zeiss, Germany) including an Axiovert 135 microscope equipped with 100 × 1.3, 40 × 1.3 (both oil immersion type) and 10 × 0.3 plan neofluor objectives. The two internal helium-neon lasers (543 and 633 nm) were used as the excitation source for the Cy3—(at 543 nm) or Cy5—(at 633 nm) labelled oligonucleotide probes. Fluorescence of the FITC dye was detected with an external argon laser (488 nm). After in situ hybridisation with rRNA-targeted oligonucleotide probes, an anti-fading agent AF1 solution (Citifluor Ltd., London, United Kingdom) was distributed onto the slides before analysis with the CLSM. The FITC-stained carrier particles were cut into halves, immersed into a phosphate buffer solution (PBS), and the peripheral biofilm along the diameter was analysed with the CLSM. All image processing was carried out with the Zeiss software package.

4. Results and discussion

Two experimental runs were made with different sources of inoculum. In the first one, a biofilm sample was taken from a pilot-scale (17 m^3) sequencing batch biofilm reactor in Ingolstadt, Germany [13]. The plant had been operated with biological phosphorus removal for 4 months. In the second run, a biofilm sample was taken from a bench-scale (20 L) sequencing batch biofilm reactor that had been operated for 2 yr. Both of these sequencing batch biofilm reactors (SBBR) applied oxygen as the main electron acceptor in the EBPR process, whereby initially only a fraction of the PAOs could denitrify. The biofilm samples were transferred to a lab-scale reactor with alternating anaerobic and anoxic conditions.

4.1. Phosphate removal activity

Figs. 1 and 2 show the phosphate outlet concentrations during the two experimental runs. Each peak on

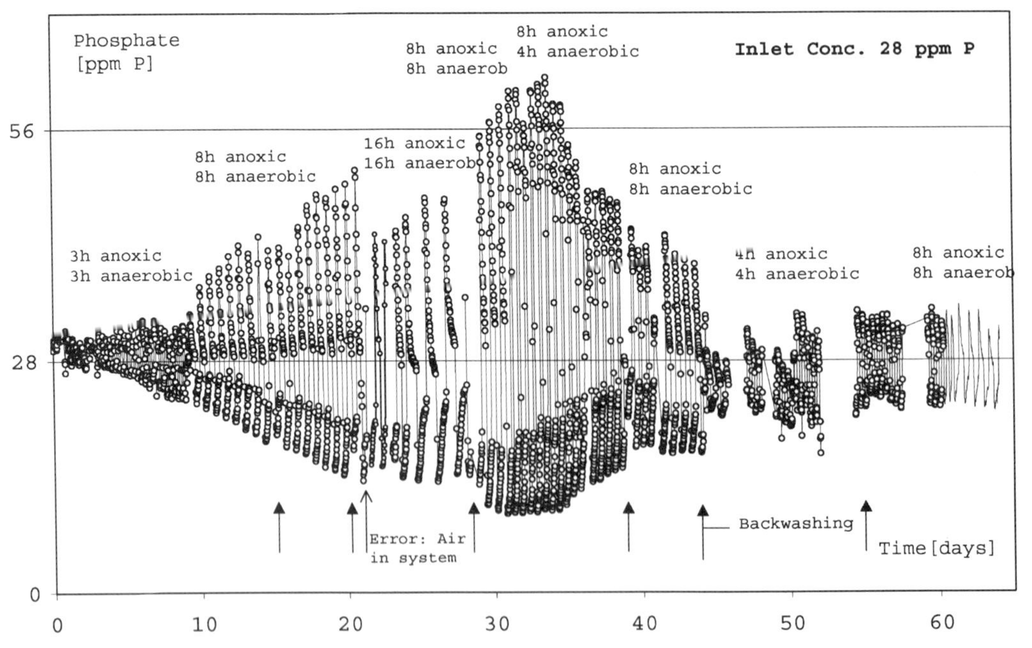

Fig. 1. Phosphate outlet concentrations during the first experimental run. The inlet concentration was constant (28 ppm P). Each peak on the curve identifies one anaerobic phase, and each valley identifies one anoxic phase. The area between the outlet concentration curve and the inlet concentration (28 ppm P) during anaerobic phases equals the amount of phosphate released from the biofilm. The area between the inlet concentration (28 ppm P) and outlet concentration curve during anoxic phases equals the amount of phosphate taken up by the biofilm. Anaerobic and anoxic phase lengths were changed during the period as indicated on the figure. Maximum phosphate removal activity occurred around day 32.

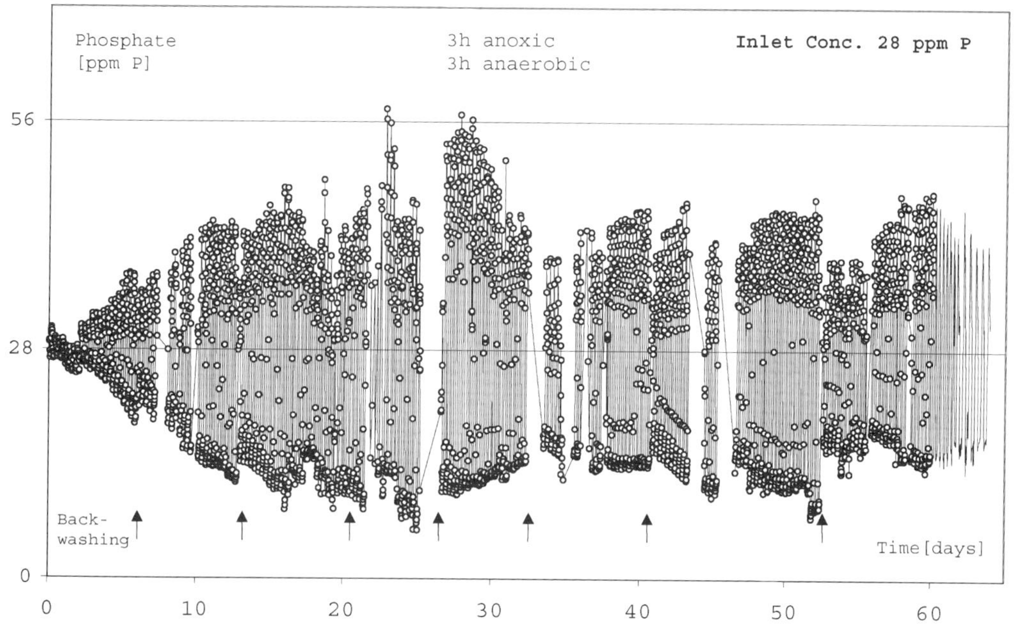

Fig. 2. Phosphate outlet concentrations during the second experimental run. The inlet concentration was constant (28 ppm P). Each peak on the curve identifies one anaerobic phase, and each valley identifies one anoxic phase (as in Fig. 1). It may be difficult to distinguish the separate cycles (first the curve goes up during the anaerobic phase, then down during the following anoxic, then up again during the next anaerobic phase, etc.) in the figure due to the very compressed curve (4 cycles per day). Anaerobic and anoxic phase lengths were not changed during the period, but kept at 3 h each. Maximum phosphate removal activity occurred around day 28.

the curve identifies one anaerobic phase. The area between the measured phosphate curve and the inlet concentration gives the amount of released phosphate during an anaerobic phase. The area between the inlet concentration and the measured curve during the anoxic phases gives the amount of phosphate taken up by the bacteria (baseline of 28 ppm P, Figs. 1 and 2).

For the first run, the phase lengths were changed about every other week as indicated in Fig. 1. The activity declined after a change of the anaerobic phase length after 32 days, and this trend was apparently non-reversible despite returning to the previous cycle configuration. The reason for this deterioration was speculated to be a shift in the microbial population caused by the changed phase length, perhaps in favour of the so-called glycogen-accumulating organisms (GAOs). Acetate was used as the only carbon source, which usually is in favour of phosphate accumulating bacteria compared to GAOs. However, due to the complications of diffusion where different compounds in the water phase outside the biofilm might penetrate the biofilm to different depths, the deeper part of the biofilm could supply a growth zone for GAOs due to the possible presence of acetate without phosphate in this region. Liu et al. [15] used a low phosphorus/acetate-COD ratio to suppress the growth of PAOs in a biological phosphate removal system and obtained an enriched culture of GAOs. For a discussion of the aspect of diffusion in a PAO biofilm see [1].

A new experimental run was started using a fresh inoculum from the aerobic bench-scale SBBR. This time, biomass samples were collected regularly and investigated with FISH. During the first 60 days the phase lengths were kept constant at 3 h. In this run, a build-up similar to the first run was seen during the first 28 days. Not much activity took place during the first few cycles upon the transfer to anoxic conditions on day 0, but hereafter, the activity steadily increased for 4 weeks followed by a deterioration from day 28 to 32. This start-up trend was similar to the observations during the start-up of the first run, day 0 to 38. However, the deterioration in the second run was not as significant as in the first run, since the activity was stabilised a few days after the peak activity and remained at a stable level from day 32 onwards. In this run, the operating conditions had not been changed, whereby the deterioration could not be explained in the way first assumed for the previous run. Furthermore, the loss of activity in the first run by the end of the period may have been due to a very rough backwash on day 45 where a lot of biomass was lost. A similar but less pronounced effect of a backwash was seen for the second run on day 52. For days 61 to 90, 5-h phase lengths were used and for days 91 to 120 8-h phase lengths (data not shown). The amplitude of the phosphate outlet curve increased a little by the use of longer phase lengths, and the bio-P-activity was stable.

4.2. Microbiological analysis

Fig. 3 shows examples of the aerobic and denitrifying biofilm thickness before and after backwash. The microbial colonization on the carrier surface was very heterogeneous and it was not possible to determine an average thickness. For one spot sample, the denitrifying biofilm thickness varied from 67 to 1096 μm before and from 0 to 469 μm after backwashing (Fig. 3a). The aerobic biofilm from the bench-scale reactor was more homogeneous and thinner. Before backwashing, the typical thickness was 100–200 μm, and after backwashing 0–50 μm. The aerobic biofilm community consisted of many protozoans. This was evident especially after backwashing (Fig. 3b). Most of the protozoans apparently stayed attached during backwashing, whereas bacterial cells detached.

Table 2 presents an overview of the results of the gene probe analysis. Almost all cells were visualised with the EUB338 probe that detects microorganisms within the *Bacteria* domain. The bacterial biofilm community from the aerobic bio-P reactor consisted of a high number of

Table 2
Results of the FISH analysis. +: Very few cells. + + + + +: Most cells. As indicated by comparison with the EUB338 probe and a transmission image. The major shift happened within the first two weeks after transfer from the aerobic to the denitrifying set-up. Bacteria belonging to the alpha subclass of *Proteobacteria* disappeared and characteristic round beta *Proteobacteria* clusters were replaced by single short rods identified as beta *Proteobacteria*. Characteristic small clusters of GPBHGC decreased in numbers and were gradually completely replaced by filamentous GPBHGC

Probe	Denitrifying bio-P reactor	Aerobic bio-P reactor
EUB338	+ + + + +	+ + + + +
ALF1b	+	+ + +
BET42a	+ + + +	+ + +
GAM42a	+ +	+ + +
CF319	+ +	+
HGC69a	+ + (small clusters → filaments)	+ + +
ACA23a	+	+
MNP1	+	+
MP2	+	+
RHC438	+ +	+ +
RHX851	+	+

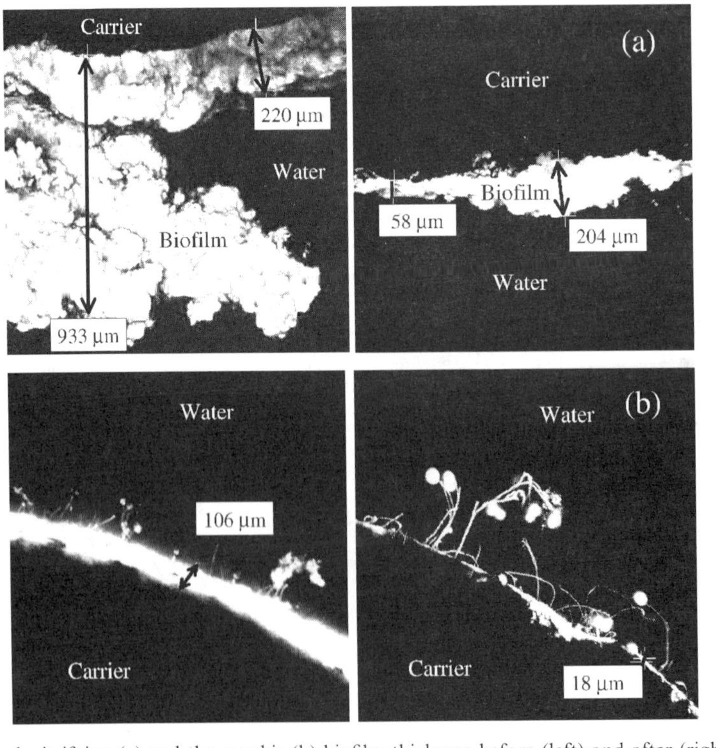

Fig. 3. Spot sample of the denitrifying (a) and the aerobic (b) biofilm thickness before (left) and after (right) backwash. The biofilm was stained with FITC which detects cells and EPS. Different carriers were used for the investigations before and after backwash.

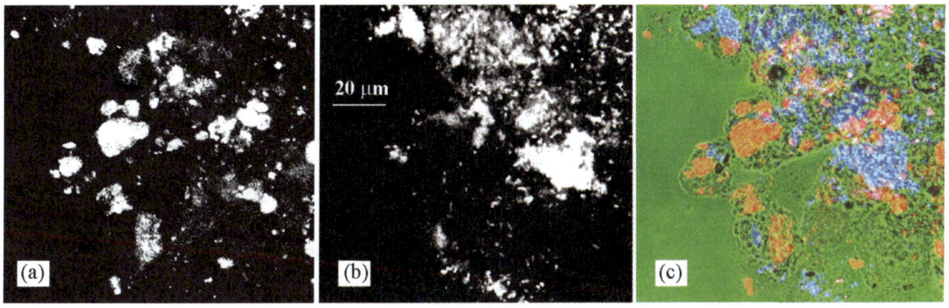

Fig. 4. FISH of aerobic biofilm samples (Day 41) with a Cy3-labeled BET42a probe (a) and a Cy5-labeled GAM42a probe (b). The microphotographs (a), (b) and a transmission image (green) were superimposed (c). The beta *Proteobacteria* clusters (orange signals) were often surrounded by gamma *Proteobacteria* clusters (blue) that appeared to "fill in" the space between the beta *Proteobacteria* clusters. The images are projections of different *xy*-sections. This causes an apparent overlap (pink colour) between some signals of the BET42a and the GAM42a probes, effected by bacteria sitting on top of one other.

alpha, beta, gamma *Proteobacteria* and Gram-positive bacteria with a high DNA G+C content (GPBHGC) which were found at a similar frequency. Many characteristic round clusters belonging to the beta subclass of *Proteobacteria* were determined in the aerobic biofilm. These clusters were often surrounded by gamma *Proteobacteria* clusters that appeared to "fill in" space between the beta *Proteobacteria* cell clusters (Fig. 4). Biofilm from the same reactor was previously investigated by Gieseke et al. [16]. These authors also observed characteristic beta *Proteobacteria* clusters and the cells within gave positive signals when using a probe for *Nitrospira* (a nitrifying bacterial genus). No significant changes were observed in the aerobic biofilm population during the sampling period, which indicated that any observed changes in the anoxic lab-scale reactor were caused by the changed environment. A noticeable shift in the denitrifying population was determined within the first 2 weeks of start-up and no change in the population was observed around the time of the activity decline after 1 month of operation. Almost no bacteria belonging to the alpha subclass of *Proteobacteria* remained in the denitrifying biofilm (Fig. 5). The beta and gamma *Proteobacteria* clusters occurring in the aerobic biofilm became less abundant in the denitrifying biofilm and were replaced by many short, oval rods belonging to the beta *Proteobacteria*, giving rise to a cohesive layer (Fig. 6). Within this layer some single gamma *Proteobacteria* cells were relatively evenly distributed (Fig 6). Bacteria belonging to the *Cytophaga-Flavobacteria* group occurred only in very low numbers in the aerobic biofilm and were more frequently found in the denitrifying biofilm. An interesting phenomenon was observed for the GPBHGC during the experimental period. The small GPBHGC clusters initially appearing were gradually completely replaced by filamentous GPBHGC during the 4-month experimental period (Fig. 7). These filaments were situated inside flocs and looked like a "floc-skeleton" in that they often followed the floc boundaries in addition to making up a web in the floc. Other filaments extended from the flocs and were detected with the BET42a probe. More filaments appeared as a function of time. Both biofilms (aerobic and denitrifying) contained only small amounts of *Acinetobacter* spp. and bacteria identified as nocardiaform actinomycetes. Also bacteria related to the *Rhodocyclus*-like clone did not seem to play a dominant role in the two investigated populations. They were detected only sporadically (RHX851 probe).

The denitrifying biofilm community was less diverse (e.g. no protozoa) than the aerobic one. This was to be expected due to the use of a single carbon source, acetate, and the use of nitrate as electron acceptor instead of oxygen. For example, nitrifying bacteria could not survive in the anoxic lab-scale reactor. The consistency of the denitrifying biofilm was different (very slimy) from the aerobic biofilm, and it is likely that more EPS was produced in this biofilm. However, since no characterisation of the EPS was performed, it is not clear whether the slimy appearance was caused by a higher quantity of EPS or perhaps a different EPS composition.

The fact that no significant change in the microbial population could be verified with the applied set of gene probes around the time of the peak activity after one month of operating the anoxic lab-scale reactor could require alternative explanations of the observed decrease in bio-P activity. One hypothesis is a change in the biofilm structure, e.g. related to the EPS production. A reduced biofilm-specific diffusion coefficient (i.e. reduced penetration of the film) could account for the lower activity level. However, it should be stressed that the study applied mainly broad phylogenetic probes and only a few genus-specific probes. Hence, despite the fact that no significant changes in the microbial population were detected during the time of the activity decline, it

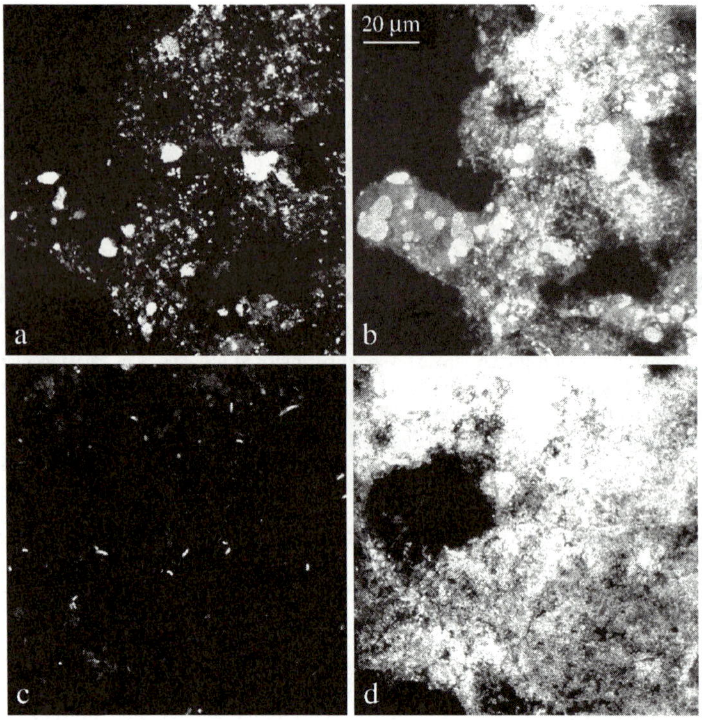

Fig. 5. FISH with a Cy3-labeled ALF1b (alpha subclass of *Proteobacteria*) and a Cy5-labeled EUB338 (*Bacteria* domain) probe. The upper half shows images of the aerobic biomass 2 weeks into the sampling period, and the lower half shows images two weeks after start-up of the denitrifying biofilm in the lab-scale reactor. Almost all of the alpha bacteria disappeared within the first two weeks following transfer of the biomass to the denitrifying setup. (a) ALF1b, aerobic sample. (b) EUB338, aerobic sample. (c) ALF1b, denitrifying sample. (d) EUB338, denitrifying sample.

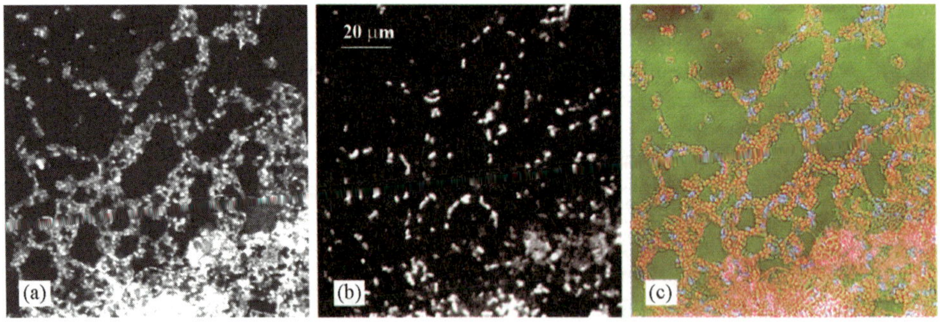

Fig. 6. FISH of denitrifying biofilm samples on day 52 with a Cy3-labeled BET42a probe (a) and a Cy5-labeled GAM42a probe (b). The microphotographs (a), (b) and a transmission image (green) were superimposed (c). Single gamma *Proteobacteria* cells (blue) appear relatively evenly distributed amongst the beta *Proteobacteria* cells (orange). The images are projections of different *xy*-sections. This causes an apparent overlap (pink colour) between some signals of the BET42a and the GAM42a probes inside the dense flocs (bottom of the pictures), effected by bacteria sitting on top of one other.

cannot be excluded that changes possibly took place *within* one of the broad groups that were investigated. A major problem regarding biological phosphate removal is the lack of knowledge of the specific organism(s) involved in this process. For an improved practical use of gene probe analysis in regard to phosphate removal, more research is needed regarding the organism(s) responsible for bio-P and the competing glycogen-accumulating organism(s). Development of new probes for these organisms would enhance investigations of the dominance of the two groups in relation to different operating conditions.

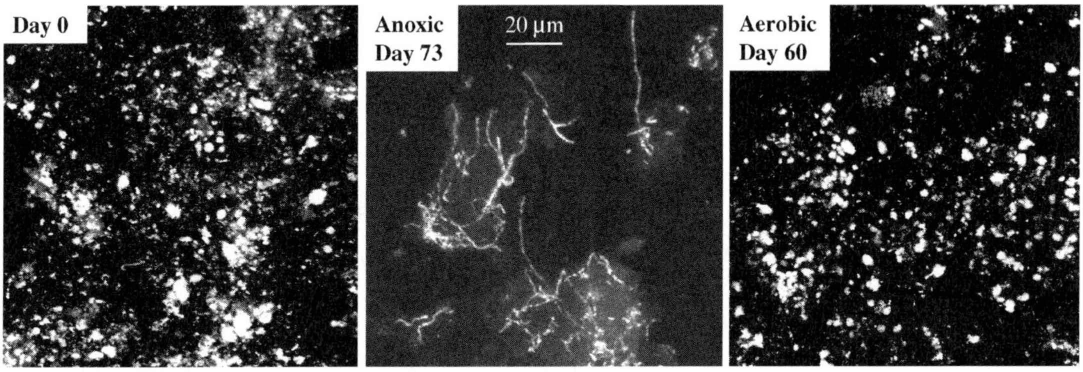

Fig. 7. FISH with a Cy3-labeled HGC69a probe on day 0 (original biofilm from the aerobic bench-scale reactor), day 73 in the anoxic lab-scale reactor, and day 60 in the aerobic bench-scale reactor. The abundance of small clusters of Gram-positive bacteria with a high DNA G+C content (GPBHGC) in the aerobic reactor decreased dramatically in numbers within the first few weeks in the anoxic reactor; they were gradually completely replaced by filamentous GPBHGC. No noticeable change was detected in the aerobic bench-scale reactor during the sampling period.

5. Conclusion

Acclimation of a phosphate-removing biofilm to nitrate instead of oxygen as terminal electron acceptor took approximately 2 weeks. FISH revealed a significant change in the microbial population during this acclimation. Apparently, the biofilm needed 1 month to adjust to a stable activity level, since a steady rise in the activity was seen in this period followed by a sudden decrease. This phenomenon was seen in two independent runs and had nothing to do with the chosen phase lengths as first assumed. This underscores the need for repeating an experiment before concluding on an observed phenomenon. FISH did not reveal any significant change in the microbial population around the time of the sudden activity decrease. However, due to the application of probes for mainly larger phylogenetic groups, it cannot be excluded that changes might have taken place within one of the analysed groups. More filaments developed in the denitrifying sludge over time. FISH showed these to belong to at least two different bacterial groups, GPBHGC and beta *Proteobacteria*. For an improved practical use of FISH in regard to the phosphate removal process, more research is recommended regarding the organism(s) responsible for bio-P and competing organisms—with simultaneous development of new gene probes. The combined study of microbial population changes and process performance is needed to understand the correlation between the two and avoid false conclusions based on only one of them.

Acknowledgements

We thank Michael Wagner and Natuschka Lee for help and advice regarding the gene probe analysis, and we thank Markus Schmid for advice regarding the FITC staining.

The research was funded by the EU-TMR-project BioToBio (Biological Nitrogen Removal: From *Bio*films *to Bio*reactors) and by The Research Center for Fundamental Studies of Aerobic Biological Wastewater Treatment at the Technical University of Munich (SFB 411, Deutsche Forschungsgemeinschaft).

References

[1] Falkentoft CM, Harremoës P, Mosbæk H. The significance of zonation in a denitrifying, phosphorus removing biofilm. Water Res 1999;33(15):3303–10.

[2] Mino T, Van Loosdrecht MCM, Heijnen JJ. Microbiology and biochemistry of the enhanced biological phosphate removal process. Water Res 1998;32(11):3193–207.

[3] Barker PS, Dold PL. Denitrification behaviour in biological excess phosphorus removal activated sludge systems. Water Res 1996;30(4):769–80.

[4] Wagner M, Erhardt R, Manz W, Amann R, Lemmer H, Wedi D, Schleifer K. Development of an rRNA-Targeted oligonucleotide probe specific for the genus *Acinetobacter* and its application for in situ monitoring in activated sludge. Appl Environ Microbiol 1994;60(3):792–800.

[5] Bond PL, Hugenholtz P, Keller J, Blackall L. Bacterial community structures of phosphate-removing and non-phosphate-removing activated sludges from sequencing batch reactors. Appl Environ Microbiol 1995;61(5):1910–6.

[6] Sudiana IM, Mino T, Satoh H, Matsuo T. Morphology, In situ characterization with rRNA targeted probes and respiratory quinone profiles of enhanced biological phosphorus removal sludge. Water Sci Technol 1998;38(8–9):69–76.

[7] Hiraishi A, Ueda Y, Ishihara J. Quinone profiling of bacterial communities in natural and synthetic sewage

activated sludge for enhanced phosphate removal. Appl Environ Microbiol 1998;64(3):992–8.

[8] Kawaharasaki M, Tanaka H, Kanagawa T, Nakamura K. In situ identification of polyphosphate-accumulating bacteria in activated sludge by dual staining with rRNA-targeted oligonucleotide probes and 4′,6-diamidino-2-phenylindol (DAPI) at a polyphosphate-probing concentration. Water Res 1999;1:257–65.

[9] Melasniemi H, Hernesmaa A, Pauli AS-L, Rantanen P, Salkinoja-Salonen M. Comparative analysis of biological phosphate removal (BPR) and non-BPR activated sludge bacterial communities with particular reference to *Acinetobacter*. J Ind Microbiol Biotechnol 1998;21:300–6.

[10] Hesselmann RPX, Werlen C, Hahn D, van der Meer JR, Zehnder AJB. Enrichment, phylogenetic analysis and detection of a bacterium that performs enhanced biological phosphate removal in activated sludge. System Appl Microbiol 1999;22:454–65.

[11] Crocetti GR, Hugenholtz P, Bond PL, Schuler A, Keller J, Jenkins D, Blackall LL. Identification of polyphosphate-accumulating organisms and design of 16S rRNA-directed probes for their detection and quantitation. Appl Environ Microbiol 2000;66(3):1175–82.

[12] Arcangeli J, Arvin E. A membrane de-oxygenator for the study of anoxic processes. Water Res 1995;29(9):2220–2.

[13] Arnz P, Arnold E, Wilderer P. Enhanced biological phosphorus removal in a semi full-scale SBBR. Water Sci Technol 2000;43(3):167–74.

[14] Amann RI. In-situ identification of micro-organisms by whole cell hybridization with rRNA-targeted nucleic acid probes. In: Akkerman ADL, van Elsas JD, de Bruijn FJ, editors. Molecular microbial ecology manual. Dordrecht: Kluwer Academic Publishers, 1995. pp. 1–15.

[15] Liu W-T, Mino T, Nakamura K, Matsuo T. Role of glycogen in acetate uptake and polyhydroxyalkanoate synthesis in anaerobic–aerobic activated sludge with a minimized polyphosphate content. J Ferment Bioeng 1994;77(5):535–40.

[16] Gieseke A, Arnz P, Schramm A, Amann RI, Wilderer P. Nutrient removal with a sequencing batch biofilm reactor: Process parameters and microscale investigations. Proceedings of the IAWQ Conference on Biofilm Systems. New York, USA, 1999;17–20 October.

[17] Amann RI, Binder BJ, Olson RJ, Chisholm SW, Devereux R, Stahl DA. Combination of 16S rRNA-targeted oligonucleotide probes with flow cytometry for analyzing mixed microbial populations. Appl Environ Microbiol 1990;56:1919–25.

[18] Manz W, Amman RI, Ludwig W, Wagner M, Schleifer KH. Phylogenetic oligodeoxynucleotide probes for the major subclasses of proteobacteria: problems and solutions. System Appl Microbiol 1992;15:593–600.

[19] Roller C, Wagner M, Amman RI, Ludwig W, Schleifer K-H. In situ probing of gram-positive bacteria with high DNA G+C content using 23S rRNA-targeted oligonucleotides. Microbiology 1994;140:2849–58.

[20] Schuppler M, Wagner M, Schön G, Göbel UB. In situ identification of nocardioform actinomycetes in activated sludge using fluorescent rRNA-targeted oligonucleotide probes. Microbiology 1998;144:249–59.

PERGAMON

Water Research 36 (2002) 501–509

WATER RESEARCH

www.elsevier.com/locate/watres

Simultaneous P and N removal in a sequencing batch biofilm reactor: insights from reactor- and microscale investigations

A. Gieseke[a,*], P. Arnz[b], R. Amann[a], A. Schramm[a,1]

[a] *Molecular Ecology Group, Max-Planck-Institute for Marine Microbiology, Celsiusstrasse 1, 28359 Bremen, Germany*
[b] *Institute of Water Quality Control and Waste Management, TU Munich, Am Coulombwall, D-85748 Garching, Germany*

Received 11 January 2001; received in revised form 4 April 2001; accepted 9 April 2001

Abstract

A sequencing batch biofilm reactor (SBBR) with well established enhanced biological phosphate removal (EBPR) was subjected to higher ammonium concentrations to stimulate and eventually implement simultaneous nitrification. Changes of activity and populations were investigated by a combination of online monitoring, microsensor measurements and fluorescence in situ hybridisation (FISH) of biofilm sections. Nitrification and nitrifying bacteria were always restricted to the periodically oxic biofilm surface. Both, activity and population size increased significantly with higher ammonium concentrations. Nitrification always showed a delay after the onset of aeration, most likely due to competition for oxygen by coexisting P accumulating and other heterotrophic bacteria during the initial aeration phase. This view is also supported by comparing oxygen penetration and oxygen uptake rates under low and high ammonium conditions. Therefore, simultaneous nitrification and phosphorus removal in a P removing SBBR appears to be only possible with a sufficiently long oxic period to ensure oxygen availability for nitrifiers.

Keywords: Biofilm; Enhanced biological phosphorus removal; Fluorescence in situ hybridisation; Microsensors; Nitrification; Oxygen budget; Sequencing batch biofilm reactor

1. Introduction

Biofilms bear great potential for the simultaneous and efficient removal of organic carbon and nutrients like N and P in wastewater treatment [1,2]. They are spatially heterogeneous, providing space for both, aerobic and anaerobic processes; they are well suited for nitrification, since attached growth of the slow-growing nitrifying bacteria protects them from washout; and they can be exposed to alternating anaerobic and aerobic conditions as necessary for enhanced biological phosphorus removal (EBPR). First examples of multiple nutrient removal in biofilm systems have already been demonstrated [2]. Recently, a sequencing batch biofilm reactor (SBBR) was designed for very efficient EBPR [3] and successfully scaled up to the pilot scale [4]. In each SBBR cycle, phosphorus is released during an initial anaerobic period. Subsequently, the reactor is aerated and phosphorus is taken up by phosphorus-accumulating organisms (PAOs). This results in a net uptake of P over the cycle. The implementation of nitrification into this reactor type is a challenging but desired step in saving reactor volume and costs. As nitrification and P removal both consume oxygen, organisms in such a system are potentially subjected to competition for oxygen. Nitrifying bacteria in pure culture are known to have a lower affinity for oxygen compared to heterotrophic bacteria as, e.g. the PAOs [5], which may result in problems when integrating nitrifying

*Corresponding author. Tel.: +49-421-2028-836; fax: +49-421-2028-690.

E-mail address: agieseke@mpi-bremen.de (A. Gieseke).

[1] Present address: Department of Ecological Microbiology, BITOEK, University of Bayreuth, D-95440 Bayreuth, Germany.

0043-1354/02/$ - see front matter
PII: S0043-1354(01)00232-9

activity. In contrast, recent in situ studies suggest the affinities of certain nitrifiers for oxygen to be relatively high [6]. Understanding the underlying mechanisms of coexistence and competition of the organisms involved is, therefore, essential to reliably set up nitrification in the EBPR–SBBR system. Knowledge of microscale function and structure of the biological components in the biofilm can help to adjust both processes to a high efficiency.

In this study, a lab-scale SBBR operated for EBPR was subjected to higher ammonium concentrations to stimulate and eventually implement simultaneous nitrification. The resulting changes in system performance, microbial activity and populations were investigated on the reactor scale and on the microscale. For that purpose, online monitoring of the process was combined with microsensor measurements and fluorescence in situ hybridisations (FISH) of microorganisms in biofilm samples with oligonucleotide probes. Our goals were to examine to what extent nitrification can be combined with EBPR in a SBBR, and to get insights in how the processes might coexist spatially and temporally on the micro- and reactor scale.

2. Methods

2.1. SBBR

A biofilm reactor with a fixed-bed volume of 20 L and an exchangeable water volume of 19 L was operated in a sequencing batch mode for EBPR as described before [3]. The 8 h operation cycle comprised a 20 min filling period, 160 min anaerobic recirculation, 260 min recirculation with aeration and 40 min draining (100% water exchange). By completely draining the reactor at the end of each cycle, carryover of nitrate to the following cycle was restricted to water remaining in pores and in the biofilm. Thus, competition for carbon by PAOs and denitrifying bacteria at the beginning of the cycle is minimised. Excess biomass was removed once a week by backwashing with pressurised air and water at the end of the aerated period. As a support material for the biofilm, $8 \times 8\,mm^2$ (diameter × height) plastic elements (Kaldnes, Purac, Merseburg, Germany) were used. These contain axial sheltered spaces that protect parts of the biomass from abrasion during backwashing (Fig. 1). Biofilm developed on the external and internal surfaces of the substratum elements, which in some cases filled the complete axial space of the substratum. Preceding observations on thin sections of the biomass hybridised with various fluorescently labelled oligonucleotide probes (see below) revealed no substantial difference in the microbial community structure of biofilm parts from the external surface and the surface part of biomass growing in the axial space.

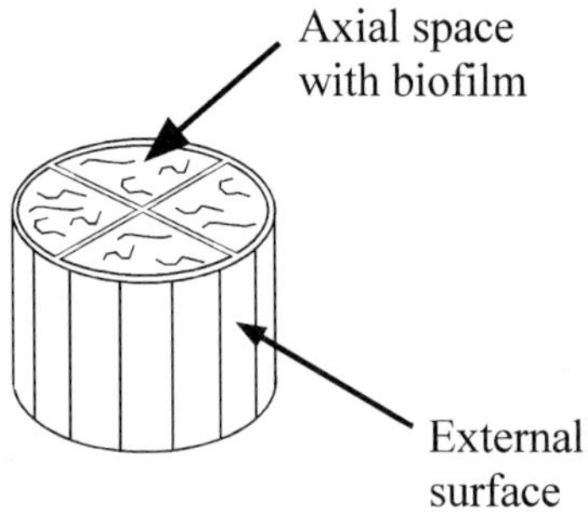

Fig. 1. Localisation of biofilm on the substratum elements.

During feeding of the reactor with a medium low in ammonium (LA operation mode), the reactor received a sterile solution containing acetate ($224\,mg\,L^{-1}$); PO_4^{3-}-P ($10\,mg\,L^{-1}$), NH_4^+-N ($13\,mg\,L^{-1}$), K^+ ($7\,mg\,L^{-1}$) and micronutrients. To enhance existing nitrifying activity, the feeding solution was enriched in ammonium and acetate was replaced by a mixture of acetate ($103\,mg\,L^{-1}$) and peptone ($200\,mg\,L^{-1}$). Thereby, the COD was maintained at the same level (i.e., $230\,mg\,COD\,L^{-1}$) while ammonium (measured as Total Kjeldahl Nitrogen; TKN) was increased to $38\,mg\,L^{-1}$ (HA operation mode). The system was allowed to adapt to the new conditions for 100 d before analyses were performed. Monitoring of phosphate and oxygen in the bulk medium was done routinely by online measurements with an oxygen electrode (Oxy 196; WTW), and with a P analyser (Phosphax inter; Dr. Lange). NO_3^--N, COD, and TKN were determined photometrically by use of standard test kits (Dr. Lange type LCK, digital photometer ISIS 6000).

2.2. Microsensor measurements

Vertical concentration profiles of oxygen, ammonium and nitrate in the biofilm were measured with microsensors [7,8]. For that purpose, substratum elements were placed in a flow chamber coupled directly to the recirculation of the SBBR, thus kept continuously under the conditions prevailing in the reactor. Measurements in biofilm grown on external surfaces or in the axial space of the substratum elements, respectively, showed no substantial differences in concentrations or shape of the profiles. All profiles presented were measured in the axial space biofilm. Measurements were performed at different times during the cycle to follow the concentration changes in the biofilm.

2.3. Calculation of theoretical and measured oxygen uptake rate

Theoretical oxygen requirements of the EBPR process were estimated based on the uptake of PO_4^{3-}-P measured during the aerobic period, and the stoichiometry of the aerobic reactions of EBPR [9]. These were compared to apparent oxygen uptake rates (OUR) for various times during the oxic period in both operation modes. Apparent OUR were calculated from microprofiles by use of the linear gradients of the oxygen profiles through the diffusive boundary layer above the biofilm as described earlier [10].

2.4. FISH analysis of biofilms

Biofilm material from the outer surface of the substratum and aggregated biomass from the axial space was sampled during both the LA and the HA operation. Samples were fixed with paraformaldehyde, cut in vertical sections on a cryomicrotome, and immobilised on microscopic slides for FISH [11]. A hierarchical set of fluorescently labelled oligonucleotide probes was used: (i) probe EUB338 [12], specific for the domain bacteria; (ii) probe HGC1351 [13], specific for gram positive bacteria with a high GC content (HGCGPB); (iii) probes ALF968 [14], BET42a, GAM42a [15], and SRB385 [12], specific for most of the alpha-, beta-, gamma-, and delta-subgroups of *Proteobacteria*; (iv) probe NSO1225 [16], specific for ammonia oxidising bacteria (AOB) of the beta-subgroup of *Proteobacteria*; (v) NIT3 [17], specific for the nitrite oxidising bacteria (NOB) of the genus *Nitrobacter*; and (vi) Ntspa712 and Ntspa662 [18], specific for NOB of the phylum and genus *Nitrospira*, respectively. Hybridisation was performed according to the conditions described by Manz et al. [15]. The distribution of hybridised cells was subsequently visualised by confocal laser scanning microscopy.

3. Results and discussion

3.1. SBBR performance

During long-term LA operation, stable P removal was achieved in the SBBR system. PO_4^{3-}-P concentrations were 27 and 0.5 mg L^{-1} at the end of the non-aerated and aerated period, respectively. NH_4^+-N concentration decreased from 13 to below 0.8 mg L^{-1} while the NO_3^--N concentration increased to about 1.5 mg L^{-1} (Fig. 2A), indicating that nitrification (and a small amount of denitrification) already was present in the system. The oxygen concentration during LA operation instantly showed a constant level of 5.4 mg L^{-1} at the beginning of aeration, and after 80 min increased linearly up to 6.8 mg L^{-1} (Fig. 2C). After the system was switched to

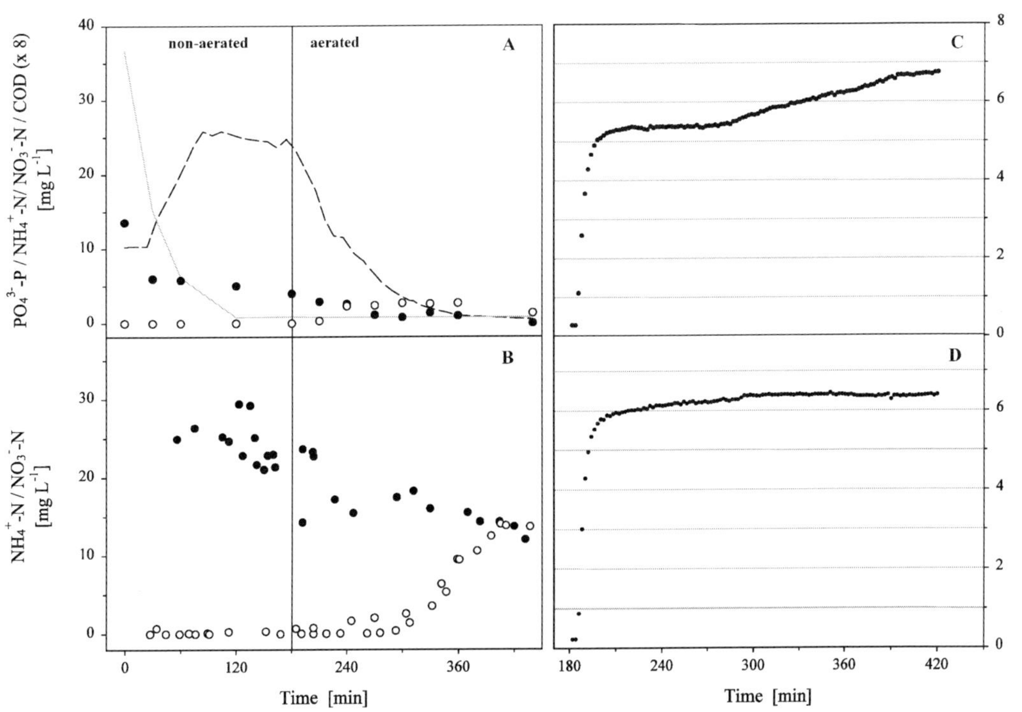

Fig. 2. Bulk concentration of PO_4^{3-}-P, COD, NH_4^+-N, and NO_3^--N during LA operation cycle (A), and NH_4^+-N and NO_3^--N concentrations as measured with microsensors in the bulk during HA operation (B), and the appropriate bulk concentrations of oxygen during LA operation (C), and HA operation (D).

HA operation, a steady state was reached within a couple of days according to bulk data. For both, the PO_4^{3-}-P and COD removal the same efficiency as under LA operation was reached (data not shown). However, NH_4^+-N concentrations now decreased from 40 to 11 mg L^{-1}, while NO_3^--N concentrations increased to 14 mg L^{-1} (Fig. 2B). In contrast to LA operation, the oxygen concentration remained constant at a level of 6.3 mg L^{-1} throughout the aerated period (Fig. 2D). No significant effect of backwashing on the bulk concentrations and, thus, on system performance could be detected in both the LA and HA operation mode.

3.2. Microprofiles

Microprofiles and online monitoring data originated from various treatment cycles between 3 and 7 days after backwashing, and a certain variability occurred between cycles. The data shown represent typical observations.

In both operation modes, oxygen was initially limited to a penetration depth of 200–250 μm during aeration. Between 80 and 140 min after start of aeration, a deeper penetration of oxygen up to a depth of 1.5 mm occurred during LA operation, coinciding with the increase of oxygen in the bulk liquid (Fig. 3). Although the absolute values differ between measurements with the online electrode and the microsensor due to different stirring sensitivity and microbial growth on the online electrode, the pattern of concentration changes was the same. Both effects, the increase in bulk oxygen and penetration of oxygen during LA operation, are in accordance with results of a simulation of EBPR in a SBBR [3]. A transition from partial to full penetration of oxygen was suggested for a 500 μm thick biofilm during aeration. This effect was ascribed to a decrease in stored polyhydroxyalkanoates (PHA) and a subsequently reduced oxygen uptake by heterotrophic microorganisms. In contrast, the oxygen penetration depth and the bulk concentration of oxygen were constant throughout the oxic period of the reactor cycle during HA operation. An additional oxygen uptake by nitrification is likely to cause this difference in the oxygen uptake and will be discussed later.

Steep ammonium profiles (data not shown) were measured during both operation modes at the beginning of the cycle that quickly levelled off thereafter. In a previous study it was shown that the biofilm strongly adsorbed ammonium [19], explaining the strong initial decrease of ammonium after filling without any production of nitrate. In contrast to LA operation, the higher amount of ammonium was not completely converted during HA operation, indicating that nitrification was not completed during the treatment cycle. In both operation modes, a low net uptake of ammonium into the biofilm was maintained throughout the cycle. However, no distinct zones of ammonium consumption

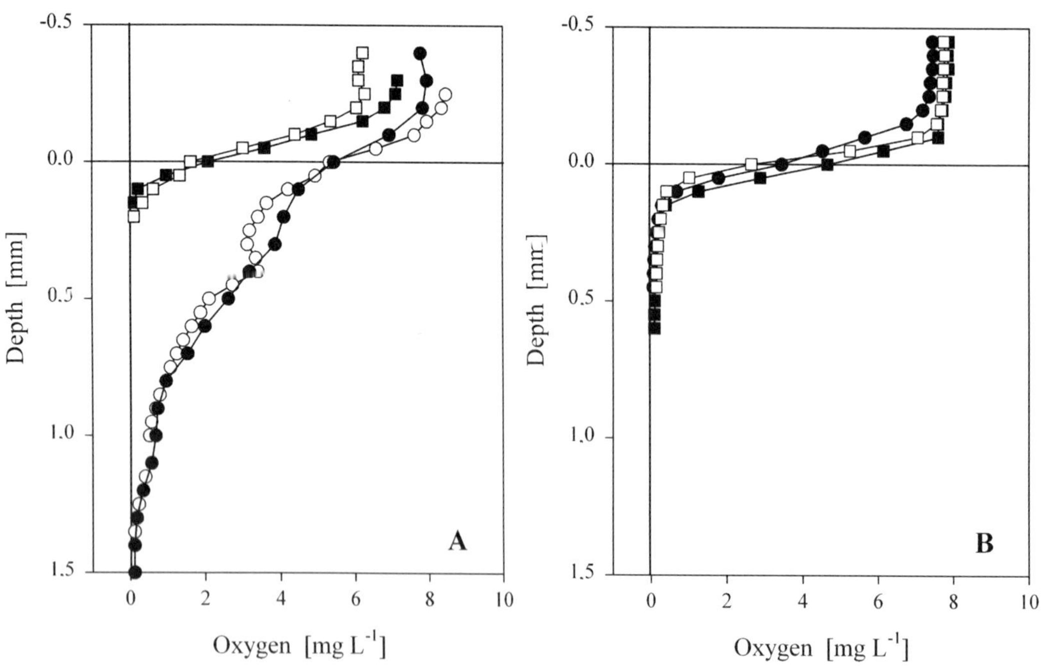

Fig. 3. Vertical concentration profiles of oxygen in the biofilm at different times of the reactor cycle during LA operation (A) and HA operation (B). Profiles were measured at 220 min (i.e. 40 min after start of aeration; □), 260 min (+80 min; ■), 320 min (+140 min; ○), and 360–370 min (+180 to +190 min; ●).

could be determined from the microprofiles. Assuming nitrification rates in the same order of magnitude as that found in other systems by use of microsensor techniques, the prevailing concentrations of ammonium were too high to support steep gradients and distinct zones of consumption.

The first occurrence of nitrate in the biofilm could be detected after approximately 60 min of aeration, and nitrate production was restricted to the oxic surface layer, i.e. to a depth of a few hundred μm. The delay in nitrification and its vertical distribution were the same in LA and HA operation mode. Nitrate production, however, was much higher in the latter (Figs. 2 and 4). Ammonium uptake without net nitrate production in the aerobic period has been demonstrated for a moving bed biofilm reactor combining nitrification and, removal [2]. It has been argued that nitrifier growth was favoured at the surface, whereas most PAOs grew in the inner parts of the biofilm and showed denitrifying activity, thereby consuming the nitrate. In our system, however, no nitrification took place in any biofilm layer at the beginning of the aeration according to the nitrate microprofiles. This can be explained by an initial competitive advantage of the heterotrophic bacteria in oxygen uptake. The putative heterotrophs might be PAOs and probably also glycogen-accumulating bacteria (GAOs [20]). Later in the cycle, the shortage of the endogenous C source and a resulting lower activity of the heterotrophs possibly enable nitrification. This hypothesis is strongly supported by the oxygen microprofiles of the second half of the aeration period. While during LA operation lower activity of the heterotrophs results in a higher oxygen penetration depth, during HA operation nitrifying activity replaces heterotroph activity at the surface in terms of oxygen consumption. As a result, the oxygen penetration depth remains constant. It can, thus, be inferred that nitrifying bacteria coexist spatially with the heterotrophic PAOs and GAOs, while their activity is separated in time.

The concentration of nitrate in the surface of the biofilm reached a value of 4.3 mg NO_3^--N L^{-1} during LA operation, and 15.8 mg NO_3^--N L^{-1} during HA operation. Therefore, nitrification can be considered to provide nitrate for either conventional denitrification or denitrifying phosphorus uptake appearing in deeper anoxic layers. This would explain the N loss observed under LA as well as HA operation. The potential carbon source for this heterotrophic process, however, remains unclear. Since the carbon source of the influent was completely sequestered in the anaerobic period, it may be assumed that storage products such as PHA were used for denitrification. It has been shown that at the beginning of the cycle acetate fully penetrated a 500 μm thick biofilm in a SBBR process [3]. Thus, carbon storage in deeper biofilm layers is possible.

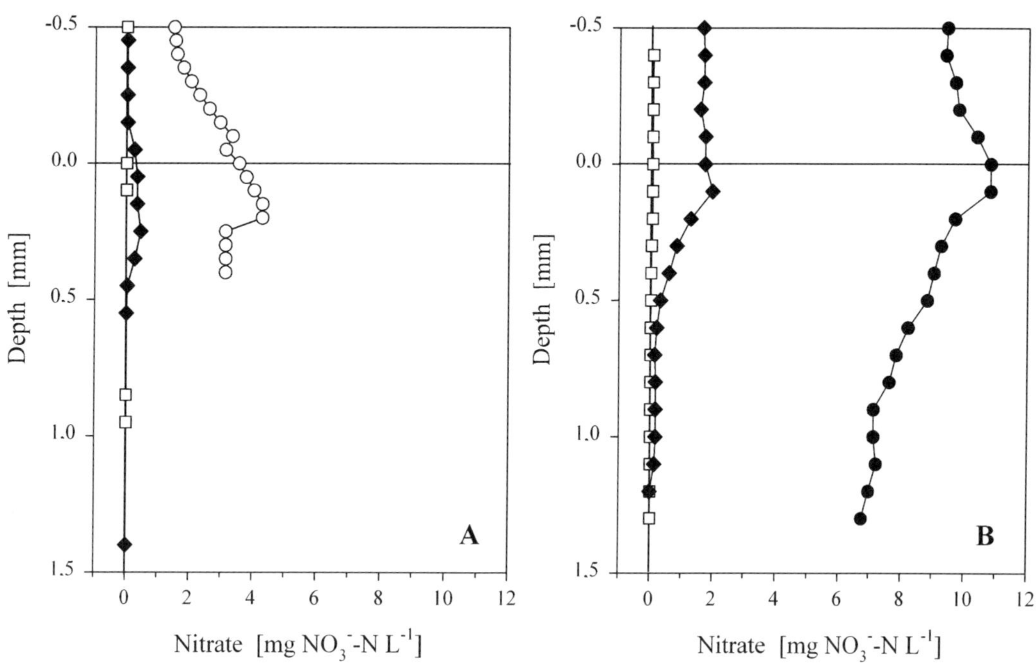

Fig. 4. Vertical nitrate microprofiles at various times of the oxic period during LA and HA operation. Profiles were measured at 220 min (i.e. 40 min after start of aeration; □), 240 min (+60 min; ◆), 320 min (+140 min; ○), and 360 min (+180 min; ●).

3.3. Theoretical and apparent OUR

The required OUR for the phosphate uptake observed for both operation modes was calculated based on a stoichiometric model for EBPR in a sequencing batch process [9]. The model assumes that four processes contribute to oxygen consumption: synthesis of polyphosphates, glycogen, and biomass, and maintenance of cell metabolism. For OUR calculation, we used a specific consumption of $0.3\,\mathrm{mol\,O_2\,mol^{-1}\,P}$ [9], an oxygen demand for growth and glycogen formation of $2.15\,\mathrm{mmol\,L^{-1}\,h^{-1}}$ [9], and a biomass composition comparable to that reported for a SBR biomass [9], i.e. a molecular weight of $26\,\mathrm{g\,mol^{-1}}$ C. The biomass concentration in the SBBR was $6.7\,\mathrm{g\,L^{-1}}$, and a maintenance demand of $4.5\times10^{-3}\,\mathrm{mol\,O_2\,mol^{-1}\,C\,h^{-1}}$ was assumed, leading to an oxygen demand of $1.16\,\mathrm{mmol\ O_2\,L^{-1}\,h^{-1}}$. In both operation modes, only the outer 300 µm of the biomass have been active and harboured potentially active cells, as shown by microsensor measurements and FISH analysis (see below). Thus, only one third of the total biomass was actively respiring. Consequently, the values for growth and glycogen formation and for maintenance have to be corrected to 0.71 and $0.38\,\mathrm{mmol\ O_2\,L^{-1}\,h^{-1}}$, respectively. This correction might underestimate the OUR of PAOs, as with LA operation oxygen penetrated to, and partly was consumed in, deeper layers during the final treatment. However, no putative PAOs were found in these layers, and oxygen consumption could not be assigned to a distinct population. The measured apparent areal uptake rates were converted to volumetric OUR by assuming an average biofilm surface of $3250\,\mathrm{cm^2\,L^{-1}}$, obtained by geometric considerations of the substratum dimensions.

Apparent OUR calculated in the beginning of the aerated period did not show differences between LA and HA operation (Fig. 5), although the initial bulk concentration of oxygen during LA operation was found to be slightly lower than with HA (Fig. 2). For both operation modes, however, the apparent OUR were found to be twice as high as the estimated oxygen requirements for the EBPR process after 80 min of aeration (Fig. 5). This suggests that either some of the parameters used in the calculation are underestimated, or that additional processes take up oxygen in the first part of the oxic period. Possibly, glycogen-accumulating organism, capable of anaerobic uptake and storage of organic carbon, were present in the system and caused this initially high oxygen uptake. However, during LA operation, OUR decreased in the second part of the aerobic period. Oxygen requirements of EBPR finally accounted for about 75% of the apparent OUR, thus indicating that at least the assumption for the maintenance demand of PAOs is realistic. Both, the decreased apparent OUR and the deeper oxygen penetration during LA operation is in accordance with predictions of an earlier model for EBPR in an SBBR [3]. In contrast, oxygen consumption remained high during HA operation. It is obvious that nitrification contributed significantly to the apparent OUR. A constant activity of nitrifiers from 240 to 450 min leading to $14\,\mathrm{mg\,L^{-1}}$ NO_3^--N during HA operation would explain an additional oxygen uptake of about $18\,\mathrm{mg\,L^{-1}\,h^{-1}}$. In contrast, during LA operation only $2\,\mathrm{mg\,L^{-1}\,h^{-1}}$ of oxygen would have been used for nitrification. Further-

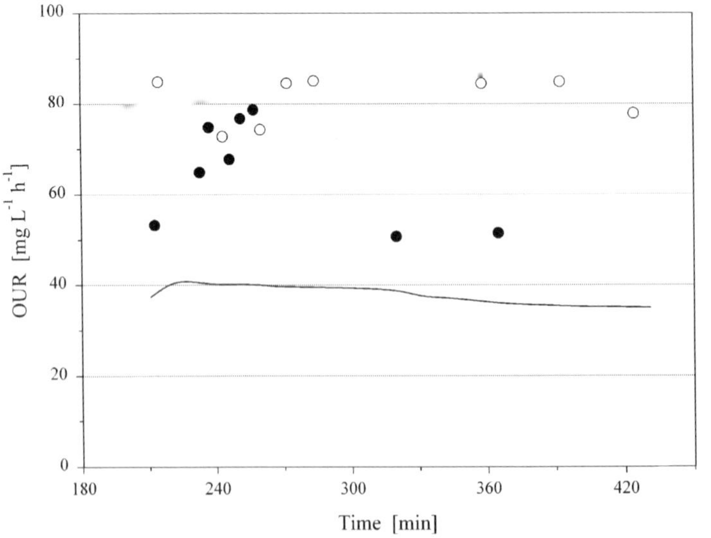

Fig. 5. Apparent OUR during LA (●) and HA operation (○), and predicted oxygen requirements for EBPR at different times of the aerobic period. For details see text.

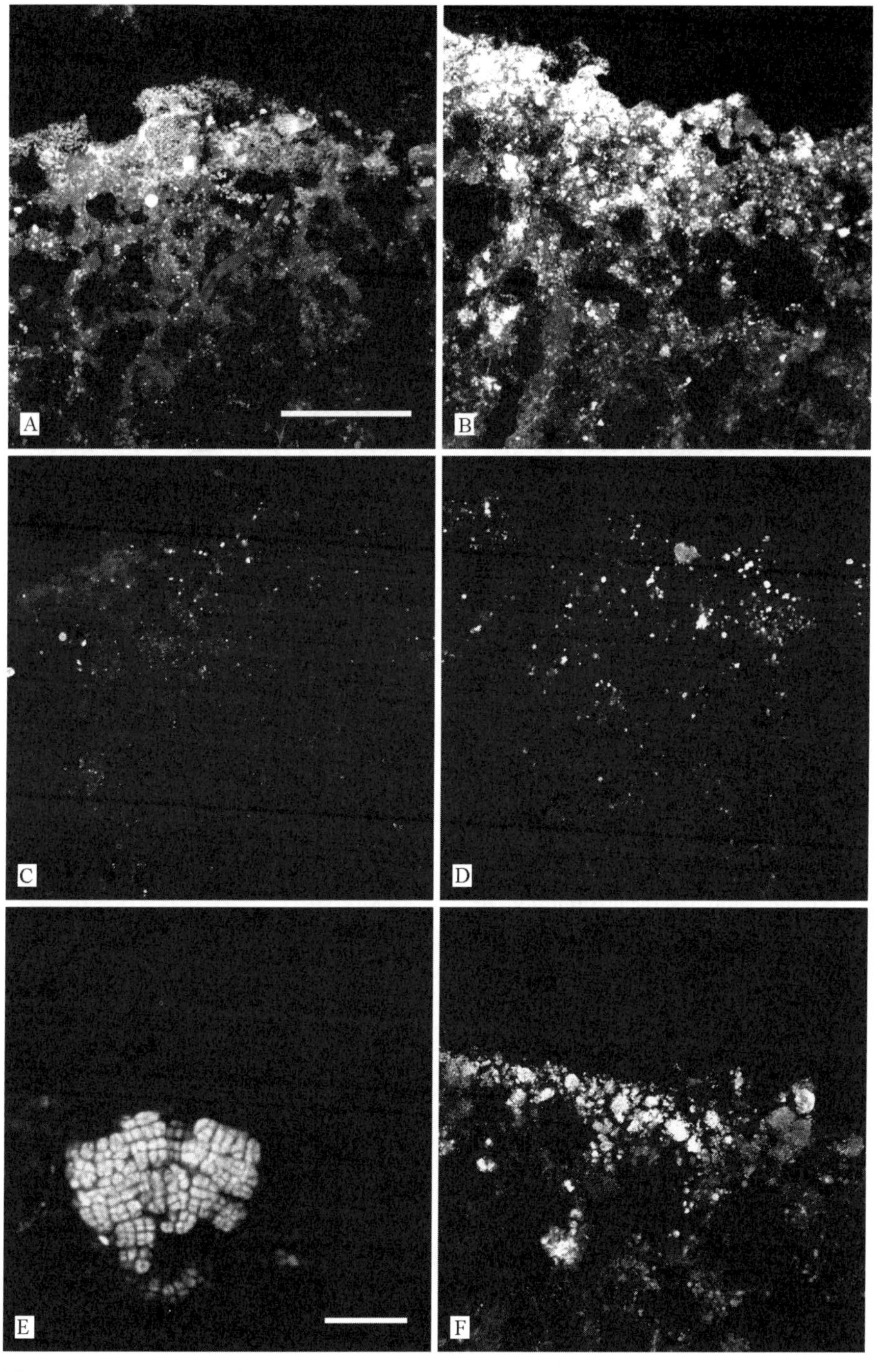

Fig. 6. Confocal fluorescent micrographs of vertical biofilm cross sections as hybridised with probes for *Bacteria* (EUB338; A and B), all nitrifying bacteria (mix of NSO1225 and Ntspa712; C and D), alpha subclass of *Proteobacteria* (ALF968; E), and Gram positive bacteria with a high GC content (HGC1351; F). A, C, and E refer to LA operation, whereas B, D and F refer to HA operation. White spots are hybridised cells or cell clusters, whereas grey areas reflect background. Scale bar is 50 μm (A–D, F) and 10 μm (E), respectively.

more, this calculation based on the accumulated NO_3^--N clearly underestimates nitrifying activity by not accounting for simultaneous denitrification.

3.4. Biofilm structure

Hybridisation with a general bacterial probe (EUB338) showed a clear stratification of the aggregated biomass during LA as well as HA operation (Fig. 6). In both cases, highest densities of bacteria could be detected in the uppermost 300 μm. Cell densities visually decreased with depth. A high number of putatively active cells were thus located in the periodically oxic biofilm layers (Figs. 6A and B), while a small amount was situated in the permanent anoxic layers of the biomass, thus possibly depending on anaerobic processes. The majority of EUB338 positive cells could be shown to belong to one of the groups described in the following.

In LA operation, HGCGPB and members of the *Proteobacteria* appearing in tetrad-like clusters were most abundant, followed by members of the beta and delta subclasses of *Proteobacteria*. As shown for other systems [21,22] the HGCGPB population might be involved in the EBPR process. The majority of this population was found in the periodically oxic surface layer (Fig. 6F). Therefore, their contribution to a denitrifying P uptake is probably of minor importance in the system. This is in agreement with the fact that no nitrate was available during the first part of aerobic period, when the P uptake was found to be high. Furthermore, this result supports the assumption for OUR calculation that only one third of the biomass is involved in P uptake. The phylogenetic affiliation of the tetrad-like clustered cells (Fig. 6E) remained unclear, as they hybridised with both the probes specific for the alpha and gamma subclass of *Proteobacteria*. Similar structures were reported from reactors with poor EBPR properties and described as the so-called "G-Bacteria" [23]. The phylogenetic affiliation of these GAOs [20] is yet unknown. The abundance of the tetrad-arranged cells visibly decreased when the system was switched to HA operation. This might indicate that these organisms have a reduced ability to utilise the substrate mixture. Despite the decline of this population, the P removal efficiency remained high, supporting the view that they play no major role in EBPR.

Beta-proteobacteria occurred in dense layers at the very surface as well as in typical spherical clusters of diameters from 4 to more than 20 μm. The latter were restricted to the biofilm surface (300 μm). Their abundance appeared to be low during LA treatment, but visibly increased after the reactor was switched to HA operation. Subsequent hybridisations with sets of probes targeting different nitrifiers showed that these clusters belong to the ammonia oxidisers of the beta subclass of *Proteobacteria* (Figs. 6C and D). Nitrite oxidisers were found to belong to the genus *Nitrospira* while *Nitrobacter* spp. could not be detected in the system. In contrast to their visual frequency, neither the composition of the nitrifying community as detected by FISH nor the vertical distribution of various populations was affected by switching to HA mode. The vertical distribution of the nitrifying bacteria in the biomass correlated to the narrow zone of nitrifying activity as detected with microsensors. Thus, nitrification was higher both in terms of the amount of detectable cells and the activity as shown by the microsensor and bulk concentration data. Their spatial coexistence with putative PAOs (and GAOs) in the biofilm supports the hypothesis of oxygen competition and a separation of both processes in time. In contrast, a separation in space, as suggested for a moving bed biofilm [2], was not found in our system.

Small irregularly shaped clusters of delta-proteobacteria were present along the entire vertical biofilm section. Since this is the most abundant group in deeper zones of the biofilm, it could be hypothesised that it participates in a conventional anoxic denitrification. However, there is not sufficient data from nitrate profiles to assign a certain zone to denitrification. Furthermore, the cell numbers appeared to be low compared to the numbers of other groups located at the surface, and the potential carbon source remains a matter of speculation.

4. Conclusions

Efficient P removal and nitrification can be achieved simultaneously in a sequencing batch biofilm reactor. Based on macro- and microscale observations, we suggest that during the oxic period P uptake and nitrification occur sequentially in time within the oxic surface layer of the biofilm. The sequential action is a result of the oxygen limitation of nitrifiers caused by competition for oxygen with heterotrophic bacteria such as the PAOs (and eventually GAOs) at the biofilm surface. The observed delay in the nitrification supports this hypothesis. Thus, besides availability of substrate, successful nitrification as combined with EBPR relies on an aerobic period that is long enough for the nitrifiers to establish.

Acknowledgements

The authors wish to thank A. Eggers, G. Eickert, and V. Hübner for providing oxygen microelectrodes, and D. de Beer for valuable comments on the manuscript. Thomas Etterer is acknowledged for providing the nitrate ionophor for the microelectrodes. This work was carried out in the framework of the Research Center for Fundamental Studies of Aerobic Biological Wastewater Treatment (SFB 411), Munich, and financially supported by the German Research Foundation (DFG) and the Max-Planck-Society.

References

[1] Watanabe Y, Okabe S, Hirata K, Masuda S. Simultaneous removal of organic materials and nitrogen by micro-aerobic biofilms. Water Sci Technol 1995;31(1):195–203.

[2] Pastorelli G, Canziani R, Pedrazzi L, Rozzi A. Phosphorus and nitrogen removal in moving-bed sequencing batch biofilm reactors. Water Sci Technol 1999;40(4–5):169–76.

[3] Morgenroth E, Wilderer PA. Controlled biomass removal—the key parameter to achieve enhanced biological phosphorous removal in biofilm systems. Water Sci Technol 1999;39(7):33–40.

[4] Arnz P, Arnold E, Wilderer PA. Enhanced biological phosphorus removal in a semi full-scale SBBR. Water Sci Technol 2001;43(3):167–174.

[5] Prosser JI. Autotrophic nitrification in bacteria. Adv Microb Physiol 1989;30:125–81.

[6] Schramm A, de Beer D, Gieseke A, Amann R. Microenvironments and distribution of nitrifying bacteria in a membrane-bound biofilm. Environ Microbiol 2000;2(6):680–6.

[7] Revsbech NP. An oxygen microelectrode with a guard cathode. Limnol Oceanogr 1989;34:474–8.

[8] de Beer D, Schramm A, Santegoeds CM, Kühl M. A nitrite microsensor for profiling environmental biofilms. Appl Environ Microbiol 1997;63(3):973–7.

[9] Smolders GJF, van der Meij J, van Loosdrecht MCM, Heijnen JJ. Stoichiometric model of the aerobic metabolism of the biological phosphorus removal process. Biotechnol Bioeng 1994;44(7):837–48.

[10] Revsbech NP, Jørgensen BB. Microelectrodes: their use in microbial ecology. Adv Microb Ecol 1986;9:293–352.

[11] Schramm A, de Beer D, Wagner M, Amann R. Identification and activities in situ of *Nitrosospira* and *Nitrospira* spp. as dominant populations in a nitrifying fluidized bed reactor. Appl Environ Microbiol 1998;64(9):3480–5.

[12] Amann RI, Binder BJ, Olson RJ, Chisholm SW, Devereux R, Stahl DA. Combination of 16S rRNA-targeted oligonucleotide probes with flow cytometry for analyzing mixed microbial populations. Appl Environ Microbiol 1990;56(6):1919–25.

[13] Erhart R. In situ analysis of microbial biocoenoses in wastewater treatment plants (In situ Analyse mikrobieller Biozönosen in Abwasserreinigungsanlagen). Thesis. Faculty for biology, chemistry and geosciences. Technical University Munich, Munich, Germany, 1997.

[14] Neef A. Application of in situ single cell identification of bacteria for population analysis in complex microbial biocoenoses (Anwendung der in situ-Einzelzell-Identifizierung von Bakterien zur Populationsanalyse in komplexen mikrobiellen Biozönosen). Thesis. Faculty for biology, chemistry and geosciences. Technical University Munich, Munich, Germany, 1997.

[15] Manz W, Amann R, Ludwig W, Wagner M, Schleifer KH. Phylogenetic oligodeoxynucleotide probes for the major subclasses of proteobacteria: problems and solutions. Syst Appl Microbiol 1992;15:593–600.

[16] Mobarry BK, Wagner M, Urbain V, Rittmann BE, Stahl DA. Phylogenetic probes for analyzing abundance and spatial organization of nitrifying bacteria. Appl Environ Microbiol 1996;62(6):2156–62.

[17] Wagner M, Rath G, Koops HP, Flood J, Amann R. In situ analysis of nitrifying bacteria in sewage treatment plants. Water Sci Technol 1996;34(1–2):237–44.

[18] Daims H, Nielsen P, Nielsen JL, Juretschko S, Wagner M. Novel Nitrospira-like bacteria as dominant nitrite-oxidizers in biofilms from wastewater treatment plants: diversity and in situ physiology. Water Sci Technol 2000;41(4-5):85–90.

[19] Wilderer PA, Arnz P, Arnold E. Application of biofilms and biofilm support materials as a temporary sink and source. Water Air Soil Pollut 2000;123(1–4):147–58.

[20] Mino T, van Loosdrecht MCM, Heijnen JJ. Microbiology and biochemistry of the enhanced biological phosphate removal process. Water Res 1998;32(11):3193–207.

[21] Wagner M, Erhart R, Manz W, Amann R, Lemmer H, Wedi D, Schleifer K-H. Development of an rRNA-targeted oligonucleotide probe specific for the genus Acinetobacter and its application for in situ monitoring in activated sludge. Appl Environ Microbiol 1994;60(3):792–800.

[22] Christensson M, Blackall LL, Welander T. Metabolic transformations and characterisation of the sludge community in an enhanced biological phosphorus removal system. Appl Microbiol Biotechol 1998;49:226–34.

[23] Cech JS, Hartman P. Competition between polyphosphate and polysaccharide accumulating bacteria in enhanced biological phosphate removal systems. Water Res 1993;7:1219–25.